VESSELS AND STATIONS OF EARTHSPACE AND THE BELT

Richard Penn

A semi-fictional work, illustrated by the author

Text copyright ©2016 Richard Penn All Rights Reserved

Interior graphics (except where noted) 2016 CC-BY Richard Penn

All illustrations may be copied under attribution share-alike terms. You are welcome to use these pictures in any way you like, as long as you mention my name. Contact me, and I'll let people know about your work.

Version 1.0 (7 July 2016)

Many thanks for technical review, by Jon "Tinker" Gardi, and Robert Zwilling, and for detailed advice from Marshall Eubanks and Hollister "Hop" David. All errors my own.

Despite appearances, this is a work of fiction. Anyone building to these plans should commit their soul to the Sucking Dark, because you will die.

Any resemblance to real persons or corporations is coincidental.

There are bound to be errors in here. If you spot some, or have better ideas about space technology, I would love to hear from you. Contact me at dickpenn@gmail.com, on the Facebook group 'Asteroid Police,' or on Twitter @RichardFPenn. I may issue revised versions of this book based on what I hear, and will announce any changes on Facebook.

Books by Richard Penn

Asteroid Police Series
1. The Dark Colony
2. Freedom at Feronia
(coming soon)
3. Traders of Arkady

Steps to Space Series
1. Spacetug Copenhagen
2. Caverns of Procellarum
(coming soon)
3. Mutiny Near Earth

Contents

1 **Introduction** 6
2 **Boats** 8
    2.1 Sally (Spacetug, 2031) 8
    2.2 Sally (Moon Lander, 2032) 11
    2.3 Scooters 14
    2.4 Tugs 18
    2.5 Buses 20
    2.6 Other boats 21
3 **Space Stations** 22
    3.1 Pharos (2031) 22
    3.3 Pharos (2035) 24
    3.4 Terpsichore Station (2051) 27
    3.5 Feronia Station (2053) 32
    3.6 Hazel Stone Station (2055) 33
4 **Utility Structures** 37
    4.1 Copenhagen Spacetug 37
    4.2 Terpsichore Hold 38
    4.3 The Cloud at Pharos 39
5 **Ships** 40
    5.1 BFPS Dancer (2051) 40
    5.2 BNS Dancer (2053) 43
    5.3 ES Shoemaker (2035) 46
    5.4 PNS Beowulf (2054) 51
    5.5 Other Ships 52
6 **Surface Stations** 54
    6.1 Moon Colonies 54
    6.2 Asteroid Colonies 57
    6.3 Martian Colonies 60
7 **Exotic Equipment** 63
    7.1 Catchers 63
    7.2 Catapults 63
    7.3 Tethers 64
    7.4 Space Elevators 64
8 **Design Considerations** 65
    8.1 Gravity 65
    8.2 Radiation 65
    8.3 Pressure 66
    8.4 Rocket Basics 66
    8.5 Propulsion 67
    8.6 Power 68
    8.7 Local Navigation 70
9 **Solar System Travel** 72
    9.1 Earthspace 72
    9.2 Travel to Mars 75
    9.3 Mars Cyclers 78
    9.4 Other Planets 79
    9.5 Transit Asteroids 80
    9.6 Colonisation Scenario 83
    9.7 Timeline 84
    9.8 What About the Stars? 84
Author's Note 85

Readers interested the technical details (who do not already know their rocket science) may prefer to read chapter 8 first.

Vessels and Stations of Earthspace and the Belt   *Richard Penn*

1 Introduction

Science fiction is littered with ships that could not possibly fly, technologies that exist only in movie special effects, and space habitats that are little more than floating wild-west towns. Space is *different*, it's *hard* to live in space. In my novels, using little more than high-school physics and the advice of scientists out on the web, I explore what it might be like to live in space using real technology we could build today, in ordinary workshops and with practical machine tools.

My thinking was inspired by a 1981 book by Jerry Pournelle. In *A Step Farther Out* (ISBN: 978-0491029414), he rigorously demolished the then-current view of ships that can dash across the Solar System in hours. The chapter *Those Pesky Belters and their Torch-ships* in particular, he makes mincemeat of the science fiction trope begun in the thirties by EE "Doc" Smith, of rockhounds zapping around the asteroid belt popping into one world or another on a day's notice. As Pournelle pointed out, those rocks are billions of kilometres[1] apart. Travelling that far in a few days requires vast planet-busting amounts of energy, tiny ships with vast fuel tanks, things that nobody will ever do. In reality, moving from one rock to another will take months at best, sometimes years, and you can only do it at a particular time, when the orbits are aligned. Pournelle's book was many years ago, but the science-fiction world did not address the issues, they just went on writing. If a film-maker could make it look convincing to an audience ignorant of space, it was real. And so Star Wars and Star Trek established a norm for stories about space travel that actually had nothing to do with space, boldly going to "another galaxy" without giving any clue to the awesome distances involved. They did not set out to show people what a real future could be like, so of course they did not achieve that goal. A few, like Arthur C Clarke, did make the effort, but even in his work a gradual accretion of the impractical crept in, with mystical and pseudo-religious guff at the end of every story.

Andy Weir, in *The Martian* (978-1785031137), is an exception where all of the science does makes sense, though I doubt NASA will ever get a mission like that out to Mars. I am sure there are other writers who make the effort.

I believe that the failure of the Apollo program to inspire, and the divergence of science fiction from science fact, have led to a hopelessness in general thinking about space, a feeling that it can't be done, or is centuries away from happening. In this book, I lay out a scenario where it *can* be done, by practical engineers within the budget of a few billionaire vanity projects, in a few decades. After that first step, it becomes self-sustaining. But only if we are willing to look at it realistically, accept the

---

[1] I unapologetically use metric units throughout. I hope this doesn't annoy US readers, but those other units *just aren't logical!*

dangers and the privations, and buckle down. Think of the crowding levels of a military submarine, of the discomfort of a tall ship in an ocean storm, not Clarke's shiny luxury stations and stewardesses with magnetic boots.

While I do believe that humans will be living in space within a few decades, I do not believe that existing national space programs will get us there. Cheap launch, of the type pioneered by Elon Musk and SpaceX, will **disrupt space**, making it possible for entrepreneurs and engineering co-ops like Copenhagen Suborbitals to build just one piece of what we need, turn it into a business, and provide a stepping stone for the next stage.

National programs support vast bureaucracies of bean counters, make bizarre decisions based on paying off their political masters, and design "missions" that spend vast efforts getting people somewhere, only to spend even more gargantuan efforts to bring them all back again. If you can get ten tonnes of something to the Moon or to Mars, you can live there and do useful work for the rest of your life. If you go there with the intention of coming back, you need another ten tonnes for a return vehicle, and a hundred tonnes of fuel to get that out there. Your project is now doing fifty times as much work, for less value. My characters don't do that; they go into space with the intention of staying. Mining, construction, building a new life in space. For them, you are not "marooned" on Mars, waiting for rescue, you are making a life there, waiting for others to join you.

People with a mission-oriented approach, and with the bureaucrat's endemic fear of failure, build everything a mission might need on Earth. Because they built it there, they can't fix it during the mission, so they have to make it super-reliable or carry backups for everything. Before they even set out, they spend years in testing everything to be absolutely sure it won't fail. Generally, during those years, their political masters will have cut the budget, so they achieve nothing from all those billions. My characters don't do that, either. They buy off the shelf or make something that works, and they go. They take raw materials and machine tools, so if a problem arises, they deal with it. The motto "**If you can't fix it, don't take it**" means that the technology level is actually quite low, the minimum needed to do the job, not the latest and greatest. As the stories make clear, sometimes your technology will kill you, but every stage in humanity's spread has meant danger to those who go first. If the alternative is to stay trapped in this gravity well, some people will make that choice.

My dearest hope is that these ideas will inspire other authors to write better fiction than mine, and ultimately make a difference to people's thinking about the future. With that in mind, feel free to use any of my works and adapt them however you wish, as long as you mention my name. But please stick to the script. The long journey times and the other limitations are there for a reason.

2 Boats

2.1 Sally (Spacetug, 2031)

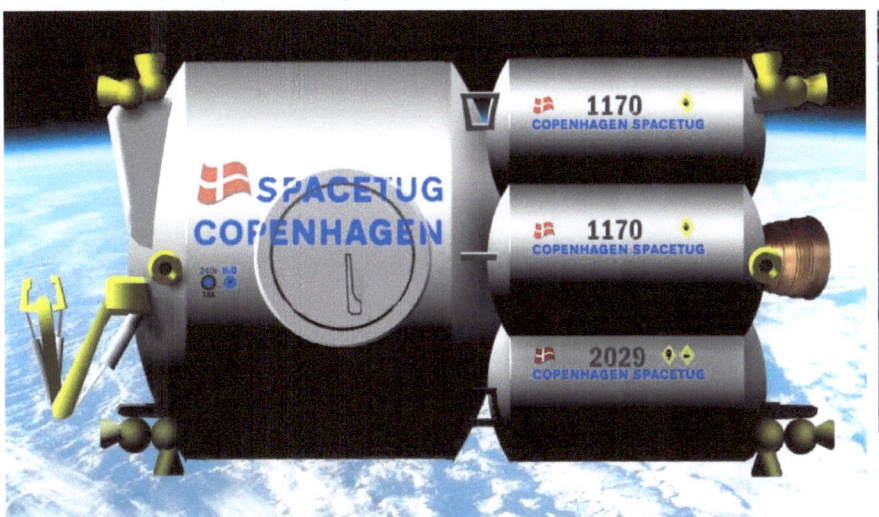

In my scenario, the first people to go to space with the intention of staying are the engineering co-op Copenhagen Spacetug, a satellite repair and reclamation business that takes over a former luxury hotel. This is their tug, named *Sally* after the late Sally Ride, the first American to go to space.

The tug is a simple plastic cylinder, with manipulators and a window on the front and a small rocket on the back. Because the

boat does not have to launch from Earth, the rocket need not be super powerful. It has enough power to recover satellites from most practical Earth orbits, including the Geosynchronous orbit (GEO) where most communication satellites operate, and the graveyard orbit where comsats retire. For many such long-distance operations, an uncrewed robot[2] vessel with similar specs (the Drone) is used, to avoid radiation exposure.

Vital to all boats used in operations around stations are the manipulators. Wherever possible, people stay

---

[2] Throughout, I use "robot" to mean a tele-operated device, not an autonomous one. People stay in control.

inside a boat and use these, rather than going out in a suit. Spacesuits are bulky and uncomfortable, exposing the wearer to more radiation and other hazards than working in a boat. And if a fault arises on a suit, all the machinery is on your back, where you can't reach it because your hands are outside. In a boat, all the equipment is in the cabin with you, so you have a much better chance of fixing it. Using the manipulators may be slightly more awkward than using spacesuit gloves, but it is far safer.

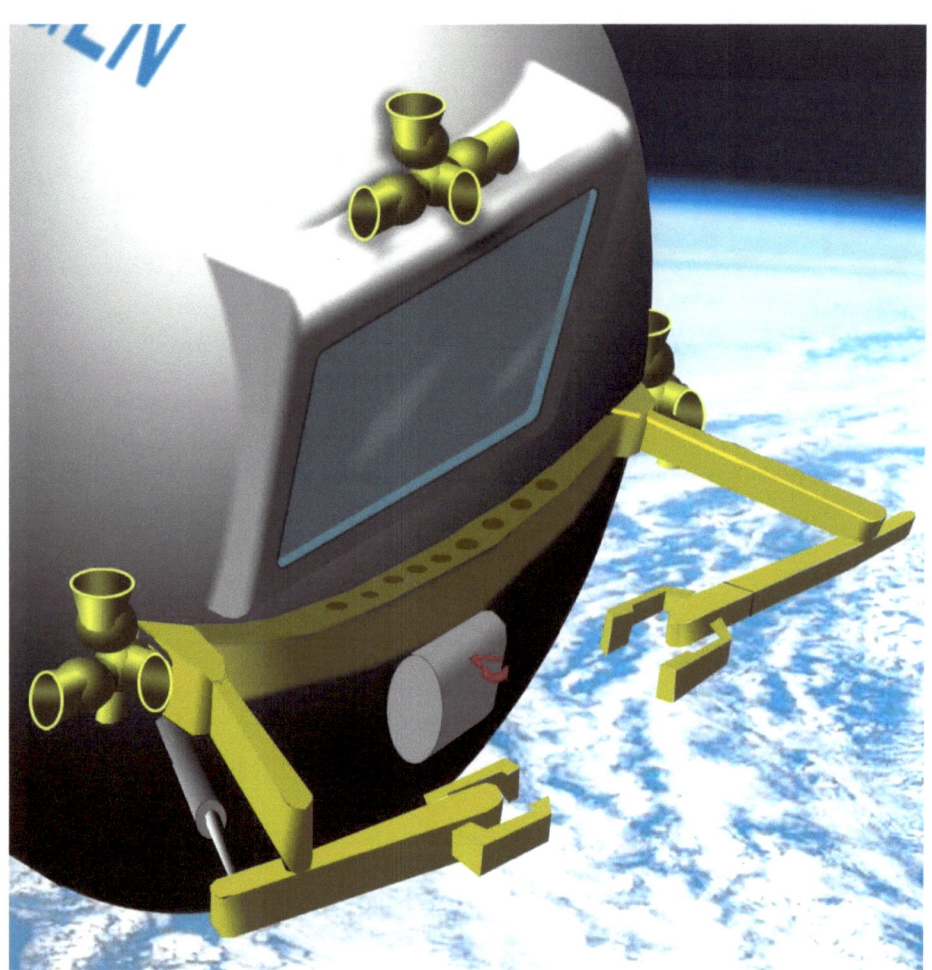

The rack beneath the front window contains power tools: drills, saws, a socket set and a winch, at minimum.

The controls of spaceboats in my stories are closely modelled on that of small submersibles operating in Earth's oceans, because the function and the requirements are very similar. All motions of the boat are controlled with the feet, leaving the hands free to operate the manipulators. The only exception is the main drive, operated by a rotary handle. This is not needed at the same time as the hands.

On the side of the boat is a hatch, capable of connecting to the station so crew can enter the tug without spacesuits. It also connects with other boats, so that crew in one boat can assist or rescue crew in another. These hatches are circular throughout my stories, like the Russian crew capsules and the Russian side of the ISS, rather than square, like the US side. A round hatch is much simpler to fly to and dock with, as the pitch attitude of the approaching boat is unimportant. The hatches are entirely mechanical and human-powered, for reliability and simplicity, built on the principle of **one hatch, one hinge, one seal**: the simplest mechanism that can do the job.

Everything in the boat runs off an internal combustion engine, referred to in the story as a 'scooter engine' (though it is probably more like a marine engine). With a power of 66kw (50hp), this is a conventional motor of the kind produced in their millions on Earth today. It is modified to run on bottled gas, the same gases that are used for the main drive. This motor provides power for all the internal systems, and acts as fuel pump for the main drive, pumping oxidant and propellant into the combustion chamber.

Each of the eight corners of the boat has a thruster unit, capable of directing steam in one of three directions. Used in sets of four, these thrusters are capable of rotating the boat in any of three directions (yaw, pitch and roll) or moving it linearly up, forward or sideways (lift, shunt and sway). The steam from the thrusters is generated using electricity from the scooter engine. Steam rockets are much less efficient than fuel-burning ones, so these operations are fairly slow. However they are safer around people in suits – the steam cools rapidly and turns to snow in the cold of space.

Vessel	**Sally (various flags)**
Commissioned	**2035 Earth (Copenhagen)**
Deployed	***Rimward* orbital hotel, later *Pharos***
Construction	**Plastic habitat, carbon-composite tanks**
Habitat	**2.5m diameter, 2.8m long, 12cm walls. 3.3 tonnes (t)**
Tanks	**6 x 1.0m diameter, 3m long, 1.2cm walls, 1.3 t empty, 13 t full**
Delta V	**Total of 5.2 km/s empty, 1.7 km/s with a 10 t load**
Acceleration	**.59 g [3] empty, .18 full**
Primary Role	**Satellite recovery/positioning in GTO, GEO or the Graveyard Orbit**
Crew	**2**
Mass	**5.3 t empty, 17 t with full tanks**

---

[3] I shall use Earth-gravity units for acceleration here. In the books, I use the metric unit (m/s²) because my characters do not like to use anything Earth-based. In fact, they often use 'millimetres,' a thousandth of that.

2.2 Sally (Moon Lander, 2032)

In *Caverns of Procellarum*, the original Spacetug is hastily adapted for landing on the Moon, by adding legs and large fuel tanks. Science fiction regularly lands big spaceships on planets with atmospheres, which is totally unrealistic and will always be so. But the moon is an exception. The gravity is relatively weak (16% of the Earth's), and there is essentially no atmosphere, so there is no need to consider air drag and heating on the descent. As a result, your ship can be any shape that will stand up.

As a tug, *Sally* had a more powerful engine than she needed just to get about in orbit. She was designed to carry an extra load of cargo, so she had enough power to carry bigger tanks instead. There is a discussion about the design in *Caverns*, where Corinne Hansen does most of the design as a student project, under Marius Kristiansen's supervision.

The ship is very challenging to land on the Moon, because her maximum acceleration with empty tanks is only 0.22g. With full tanks, she would be too heavy to take off at all. That is because the original flight was from low Earth orbit. Her total delta-V performance is over 5 km/s, burning most of the fuel in getting

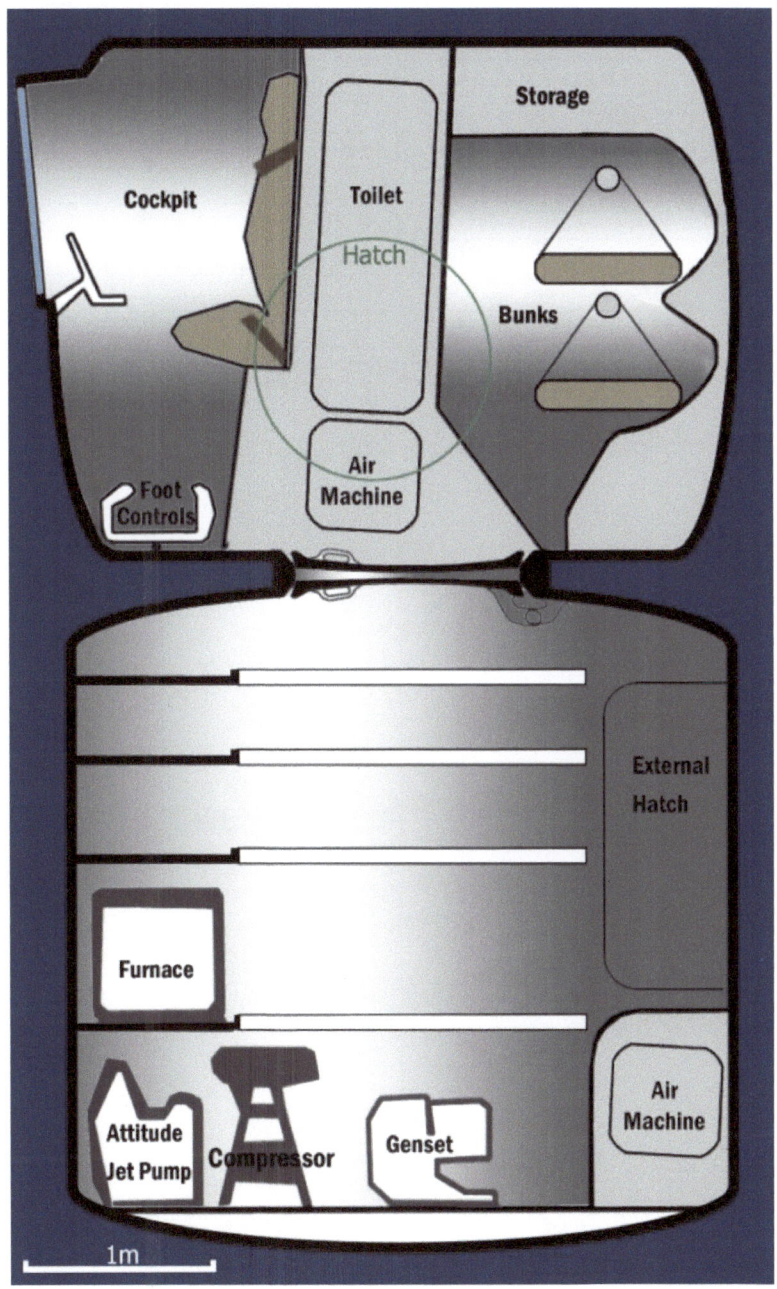

away from the Earth's gravity. The landing would have been easier if they had jettisoned two of the tanks in Lunar orbit, but they felt that would be wasteful. After arrival at the Moon, unable to lift off with full tanks, they remove two of them (useful for future ships) and proceed with hops across the surface. Her final flight from Moon to EML-1 would need those two remaining tanks to be completely full.

Vessel	**Sally (Singaporean initially, then Earthspace)**
Commissioned	**2036 at *Pharos* station, in LEO**
Deployed	***Qianshao* colony, Procellarum region, the Moon**
Construction	**Plastic habitat, carbon-composite tanks**
Habitat	**2 x 2.5m diameter, 2.8m long, 12cm walls. 6.4 tonnes**
Tanks	**4 x 2.8m diameter, 5m long, 1.2cm walls, 7.1 t empty, 102 t full**
Delta V	**Total of 6.2 km/s**
Acceleration	**.05 g full, .28 g empty**
Primary Role	**Rescue of Lunar colony, repair and recovery on surface**
Crew	**2**
Mass	**16 t empty, 10t cargo and supplies, 128 t all up**

2.3 Scooters

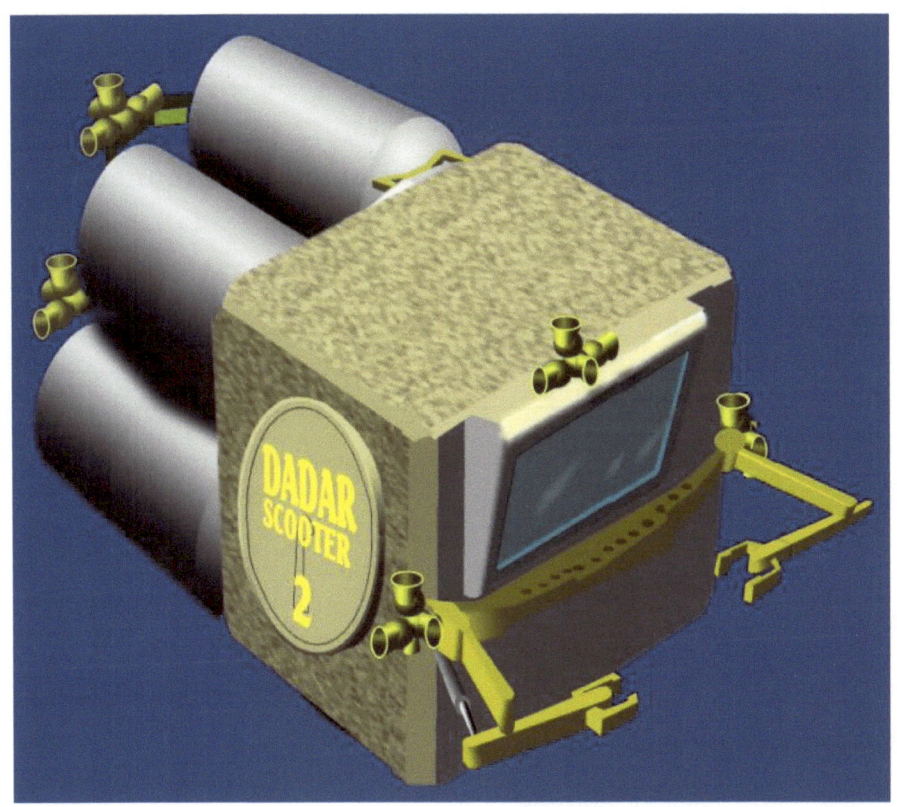

When space fiction includes something called a scooter, it is like a flying motorbike that you ride on wearing a spacesuit: essentially a rocket motor with fuel tanks attached. I do not include vehicles like this, because I think they are a bad idea. Being in a spacesuit is dangerous enough, exposing you to radiation and risking death from suit damage, without strapping a live rocket between your legs. I prefer to use small enclosed vehicles with manipulators on the outside, like a tug but smaller. I call these scooters; Arthur C Clarke called them pods, as in "open the pod bay door, HAL," but this seems an unnatural name for a vehicle.

My early scooters, from the *Steps to Space* series, are boxy. The boats are made on the Moon, from big slabs of plastic, while the rocket engines are still being imported from Earth.

Vessel	**Various scooters (numbered)**
Commissioned	**2037 Tycho, on the Moon**
Deployed	***Pharos* at EML-1 and on ES *Shoemaker*.**
Construction	**Plastic habitat, carbon-composite tanks**
Habitat	**2.4m square, 2.7m long, 12cm walls**
Tanks	**6 cylinders, 0.8m diameter x 2m long, 680kg empty, 4.8 tonnes capacity**
Delta V	**2.4 km/s**
Acceleration	**.15 g full, .27 g empty**
Primary Role	**Construction and passenger transport, local to ship**
Crew	**2, with suits if needed**
Mass	**5.7 t empty, 10.6 t with full tanks**

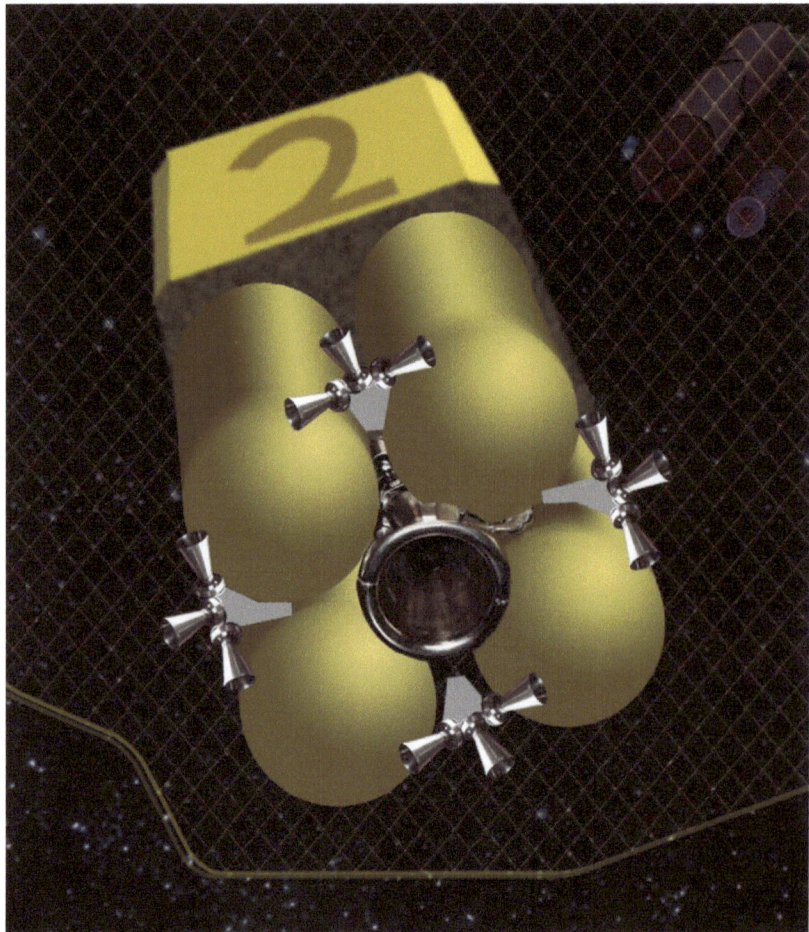

Acceleration with the main drive is similar to a small car on Earth, but the thrusters are quite weak, so turning takes a minute or more.

Fifteen years later, at the time of the *Asteroid Police* series, scooters have become a bit less primitive, though the concept is essentially the same. The body is curved, reducing the weight and making it less likely to catch on something during manoeuvres. This boat, like all in that period, uses the spherical carbon-composite fuel tanks known as "bubbles," with a better mass ratio than the old cylindrical tanks.

Vessel	**Various scooters (numbered)**
Commissioned	**2045 at Hamburg station, later at asteroid 81 Terpsichore**
Deployed	**Same**
Construction	**Plastic habitat, carbon-composite tanks**
Habitat	**2.4m square, 1.3m long**
Tanks	**4 spherical tanks, 1m diameter, 240kg empty, 1.7 tonnes capacity**
Delta V	**1.59 km/s (no cargo)**
Acceleration	**.23 g full, .35 g empty**
Primary Role	**Construction and passenger transport, local to ship or station**
Crew	**2**
Mass	**3.3 t empty, 5.0 t with full tanks**

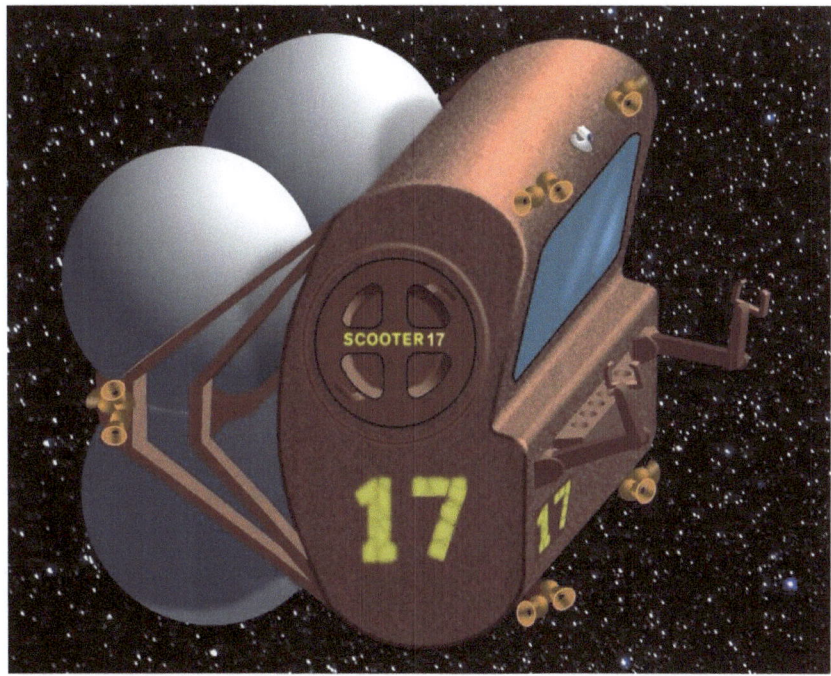

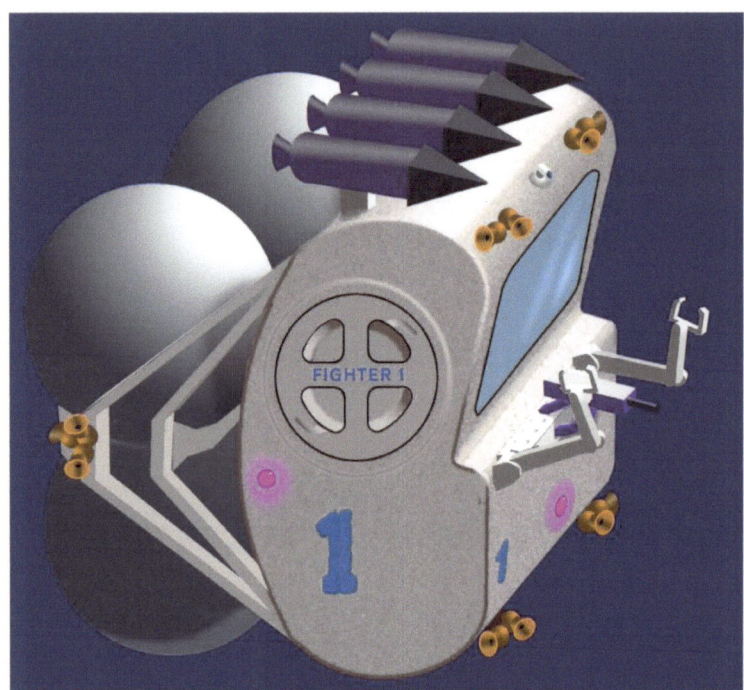

In *The Dark Colony*, these scooters are fitted with metal shrouds for reduced radar visibility, an automatic rifle, and a few rocket missiles to turn them into perky little fighting vehicles.

The "stealth" capability of the metal shroud only works in one direction, so the you can sneak up on an enemy only if you know where he is. Once in combat, it is more of an encumbrance.

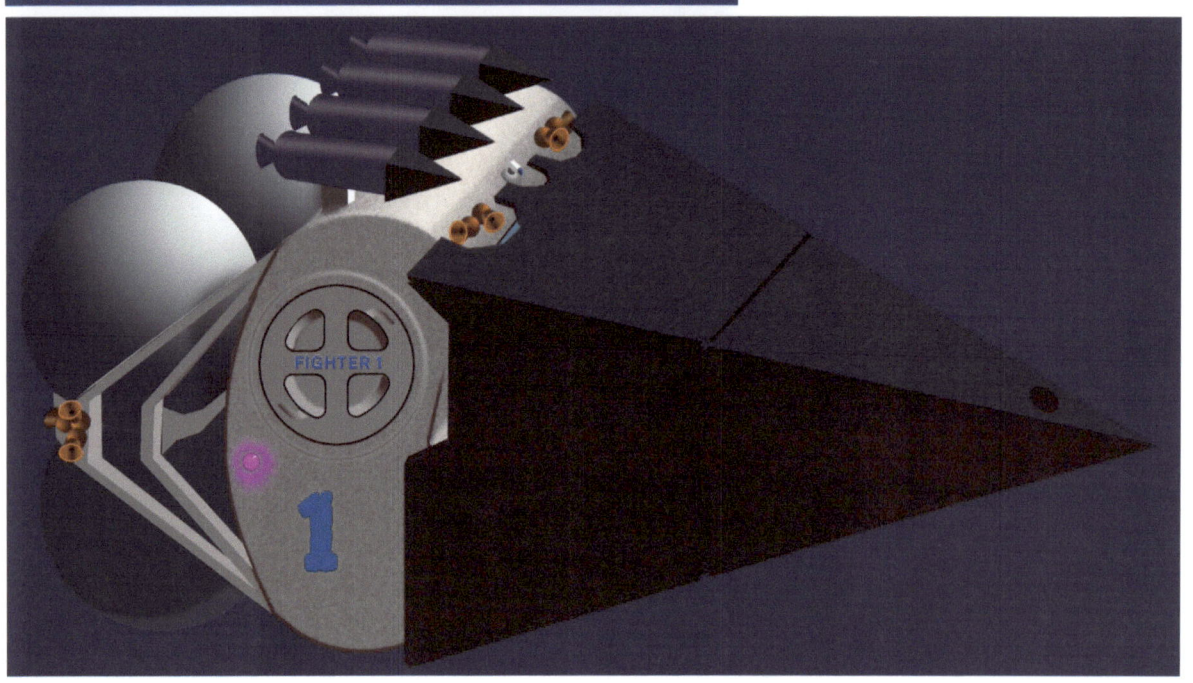

2.4 Tugs

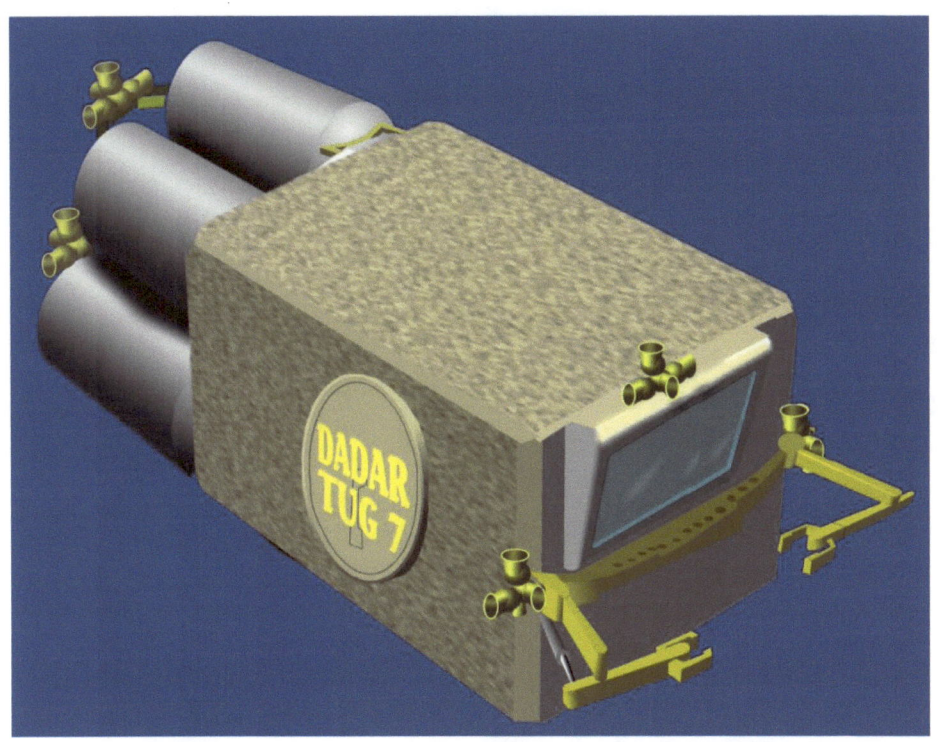

The tug is a direct descendent of Spacetug *Sally*, with more fuel tanks for extended range, and a beefier engine (with a turbo-pump in place of the gas engine) to handle bigger loads. Cargo containers are typically around 10 tonnes, and cylindrical so that they can be launched by medium-sized rockets from the Earth.

These early, box-like tugs can carry six people, or four in spacesuits. They are made from moon-plastic slabs, in the same way as the scooters.

Tugs are intended for longer-duration missions, so they have a primitive toilet. There are hatches on both sides and a tiny one-person airlock, making for more flexibility in using the tug with other boats.

Vessel	**Various tugs (numbered)**
Commissioned	**2037 Tycho, on the Moon**
Deployed	***Pharos* at EML-1 and on ES *Shoemaker*.**
Construction	**Plastic habitat, carbon-composite tanks**
Habitat	**2.4m square wide, 4.7m long**
Tanks	**6 x cylinder, 0.8m diameter, 4m long, 1.9 t empty, 15.6 t capacity**
Delta V	**3,350 m/s (no cargo), 2,130 m/s with 10 t load**
Acceleration	**.12 g full, .27 g empty, .08 g with load**
Primary Role	**Cargo, construction and passenger transport**
Crew	**2, and 2-4 passengers**
Mass	**11.4 t empty, 27 t with full tanks**

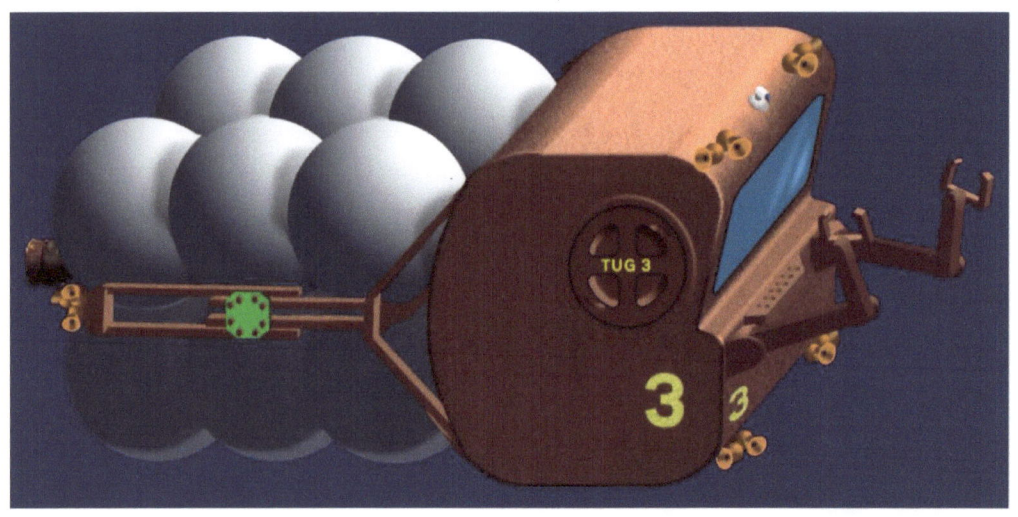

The later tugs have a more rounded profile, using the efficient spherical fuel tanks common by that time.

Out in the Belt, these tugs are used to catch and launch unpowered barges. A ten-tonne barge full or grain or other supplies is launched from Phobos on an economy orbit, taking over a year to get to 81 Terpsichore. A crew flies a tug and one or two scooters out a million kilometres or more, to catch the barge and slowly decelerate it, taking one or two days to match velocities with the station. A great adventure for the young lockhands, giving them the ultimate in privacy as well.

Vessel	**Various tugs (numbered)**
Commissioned	**2045 Hamburg station**
Deployed	**2045-50 Hamburg station, later at 81 Terpsichore**
Construction	**Plastic habitat, carbon-composite tanks**
Habitat	**2.4m square, 2.4m long**
Tanks	**12 x spherical, 1m diameter, 700 kg empty, 10 t capacity (may use more)**
Delta V	**4.1 km/s (no cargo)**
Acceleration	**.30 g full, .88 g empty, .18 with 10 tonne cargo**
Primary Role	**Construction and passenger transport, local to station. Also used to boost the station itself, for spin adjustment and hazard avoidance.**
Crew	**2**
Mass	**5.3 t empty, 15.4 t with full tanks**

2.5 Buses

When transporting larger numbers of people, an extended cabin is added to a scooter or tug. Used for 'commuter' runs, for people who work in the hold (see page 38) but live on station.

One from ES *Shoemaker*, constructed while on the voyage:

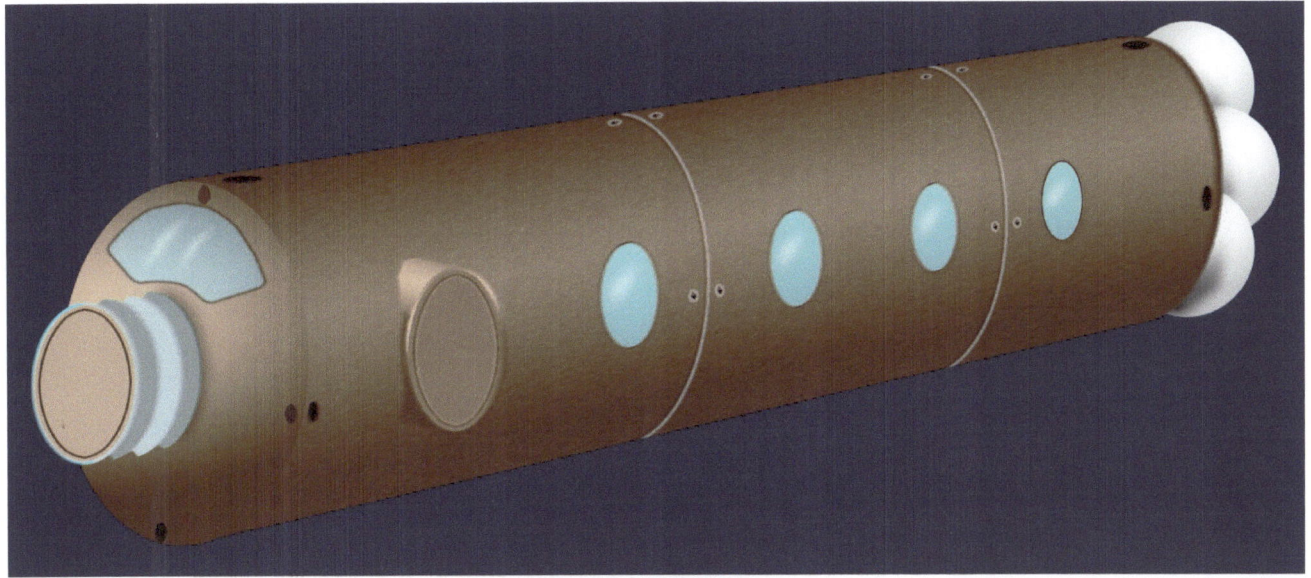

...and one from Terpsichore, used by the lockhands and mechanics to commute to the hold. The specs in both cases are similar to a scooter, but slower.

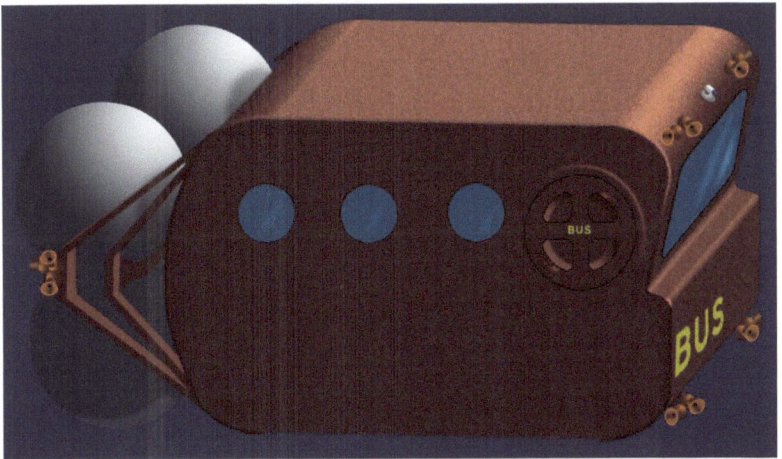

2.6 Other boats

Yet to be worked out in detail...

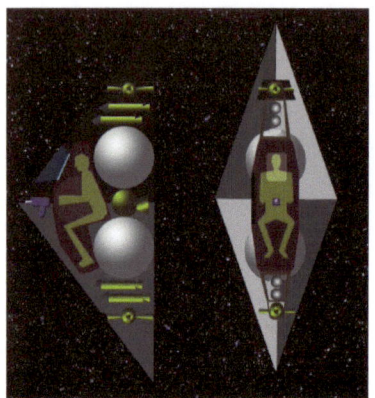

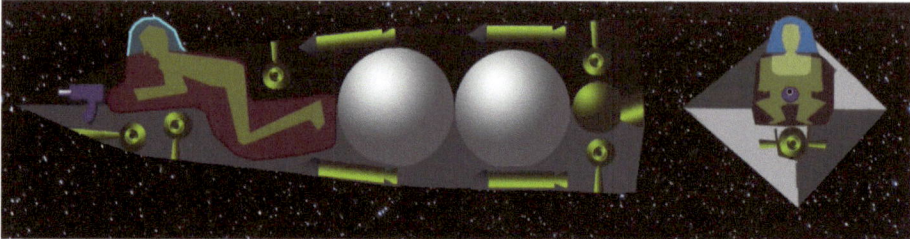

Fighting scooters, from *Traders of Arkady*

↓ Captain's Gig, also from *Traders*

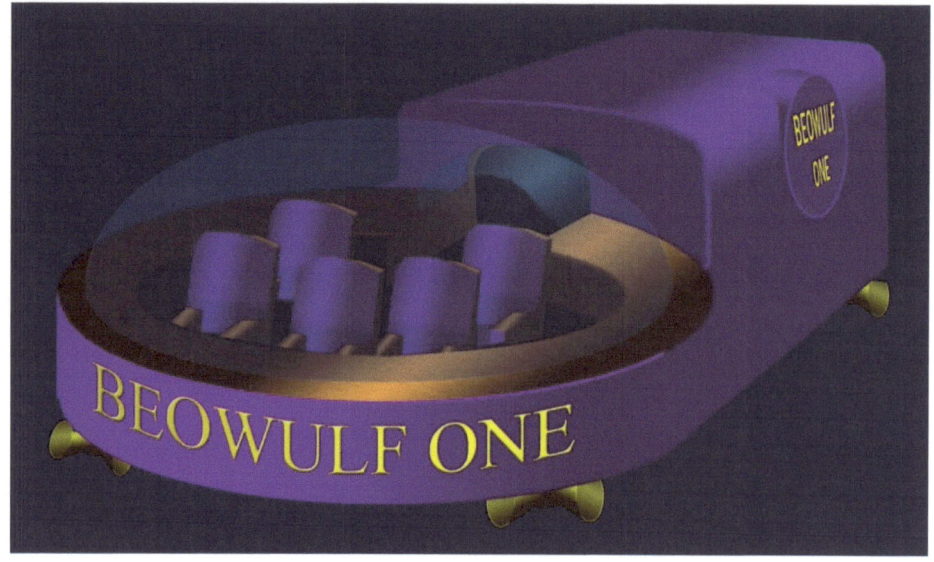

← The surface-orbit shuttle, from *The Dark Colony*

3 Space Stations

A station is a habitat (a place where people live) intended primarily to stay in one place. Of course nothing in space is truly stationary, but a station is in a fixed orbit: it is not intended to travel from one place to another. The International Space Station (ISS) is the only 2016 example of a station, orbiting in LEO. In my scenarios, stations have thrusters to adjust their orbits, increase or decrease the spin, or to dodge incoming rocks, but they do not have large main drive rockets. Add large rockets to a station, and you have turned it into a ship.

3.1 Pharos (2031)

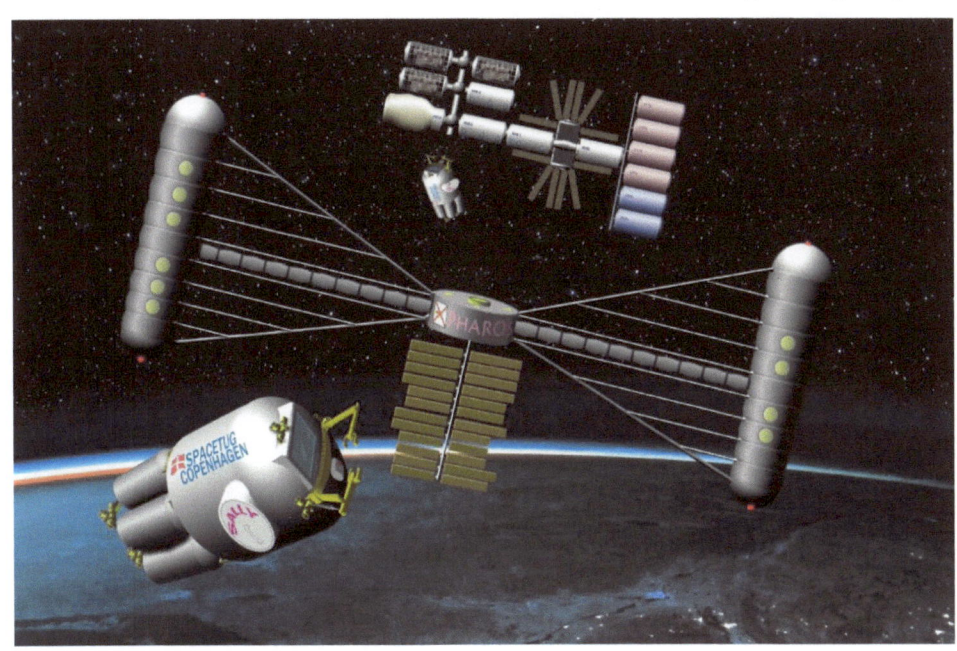

The station from *Spacetug Copenhagen* in the centre of this picture[4]. (For the small boat in the foreground, see page 8. For the utility structure in the background, see page 37.)

Pharos station provides artificial gravity by spinning on its axis. As in all my designs, the entire structure spins. I do not believe it is practical to have part of a vessel spin and another remain stationary, while providing a pressurised connection between the two.

In this case, imagine the habitat on the right (the servants' quarters and stores) are coming towards you, and the habitat on the left (the luxury hotel) is going away from you. The whole thing rotates about twice a minute, for 16.7% of Earth gravity, equivalent to the Moon.

---

[4] Background image credit: NASA/International Space Station

This station appears in *Spacetug Copenhagen* and the start of *Caverns of Procellarum*. I have no internal schematics for this station, as she was built as a luxury hotel using principles my characters do not approve of.

Vessel	**Rimward Hotel, later *Pharos* Station**
Commissioned	**2030, Earth (USA)**
Deployed	**Low Earth Orbit (later to EML-1)**
Construction	**Steel frame and plastic composite (expandable)**
Habitat modules	**5m diameter, 18.8m long, 20cm walls (2 modules)**
Rotation	**40m diameter, 25 seconds per rotation, 17% gravity**
Population	**12 as hotel, later 30**
Mass	**116 tonnes as hotel**

3.3 Pharos (2035)

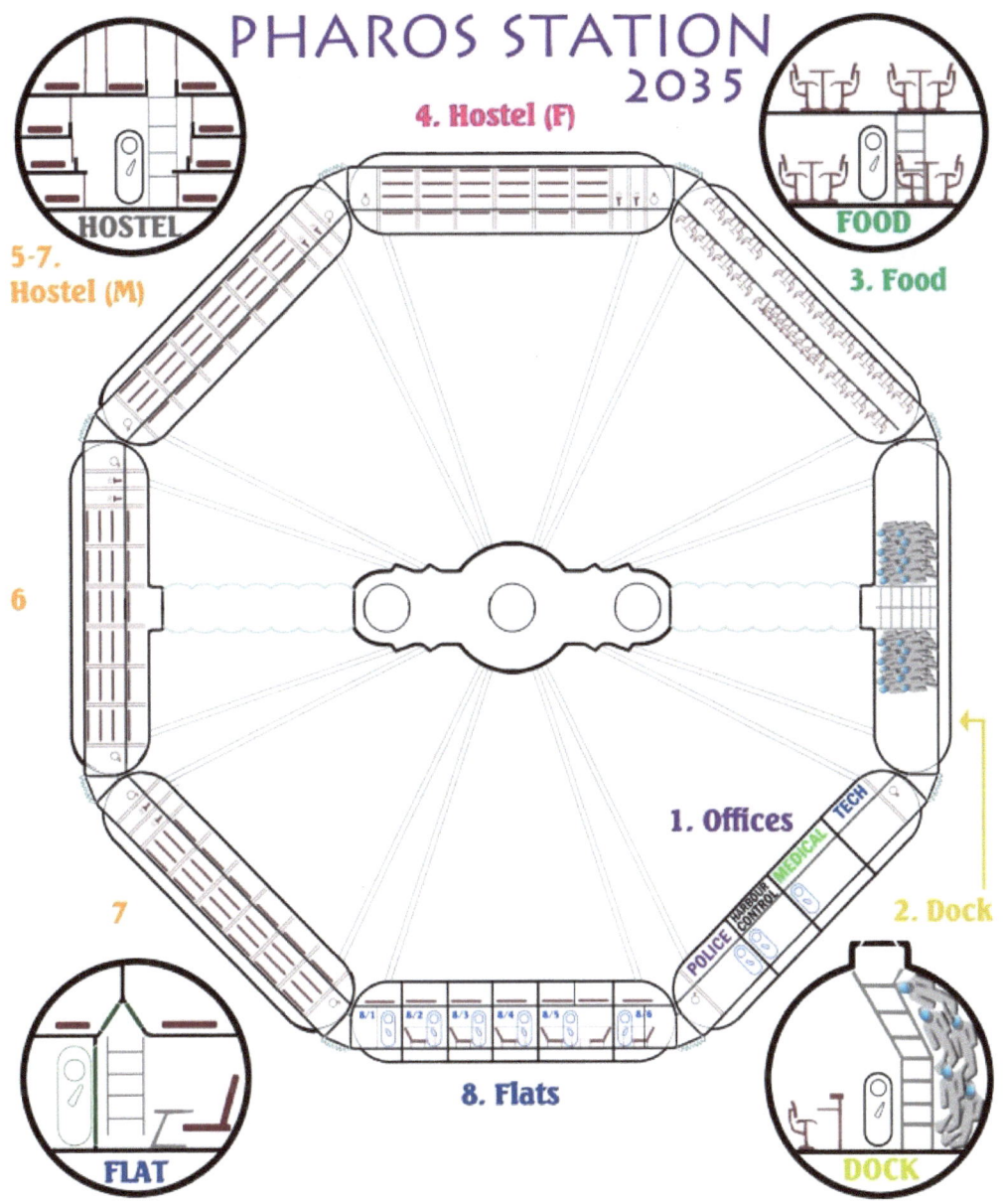

Building on from the hotel station to create a complete ring, the owners of *Pharos* move it from low-Earth orbit (LEO) to a point about nine-tenths of the way from Earth to the Moon. There is a relatively stable point there, known as Earth-Moon Lagrange Point One (EML-1). A ship or a station can slowly orbit that point, taking thirteen days for each swing, using tiny amounts of fuel to 'station-keep' – firing thrusters every few days to fix what would otherwise be an unstable orbit.

Positioned there, it is an ideal base and transfer point for the Moon, sixteen hours from anywhere on the surface, with direct line-of-sight radio or laser to talk the whole near side. A simple comsat orbiting EML-2 connects to anyone on the far side. Three days from Earth, it is a staging point, market place and emergency base for everywhere on the Moon. In my stories,

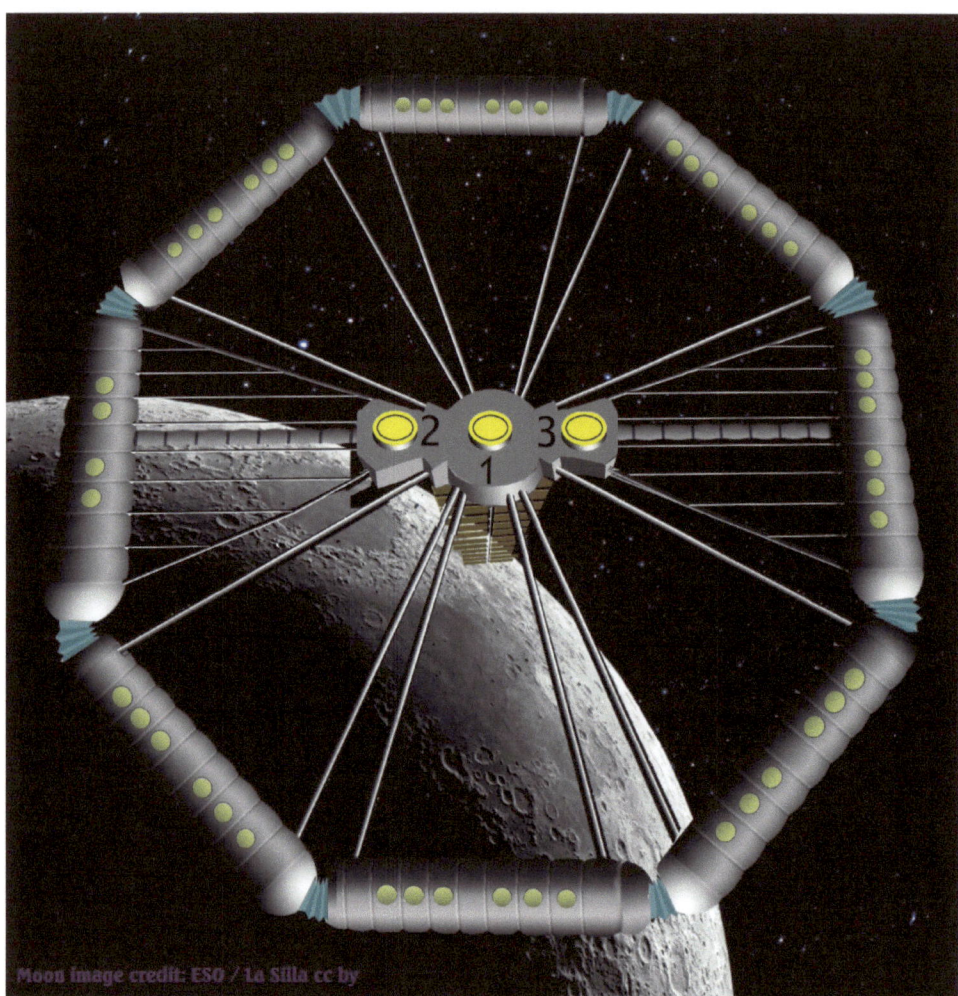

it is the capital of a sovereign state called Earthspace, which establishes much of the legal and political framework for living in space, along more-or-less democratic lines.

This station has not appeared yet in my published fiction. It is the setting for much of the upcoming *Mutiny Near Earth*, along with ES *Shoemaker* (page 44).

Apart from *Shoemaker* and a Mars ship named *Podkayne*, there are no interplanetary ships in this period, but there are regular ferries to LEO and various points on the surface of the Moon, and numerous small boats and cargo barges buzzing around cis-Lunar space.

The interior is very crowded, especially at times when a large mission is setting off or has just returned. In all my stories, people sleep in crowded dorms or tiny cabins, but spend most of their days in common areas. These (Hab 3 above) work as restaurants at mealtimes, as schools during the day, and as social space in the evenings.

A curious anomaly: *Pharos* was built from straight lozenge-shaped habitats meant to rotate "sideways," but once additional units were added, it has to rotate in its own plane, like a wheel. As a result, although the corridors look flat, the "gravity" is actually sloped. As you walk toward the end of a unit, you feel gravity up to fifteen degrees out of vertical, as if you were walking into a headwind. I assume people get used to this, but like the Lunar gravity, it is a queasy experience for newcomers.

A limitation: because the whole station rotates, including the hub, docking a boat at one of the ports is quite a challenge. I always have the computer do it, in stories. The middle port would be reasonable, but the outer ones are orbiting as well as rotating.

Vessel	**Pharos Station**
Commissioned	**2032 Earthspace (modules built on the Moon)**
Deployed	**A 13-day halo orbit of Earth-Moon Lagrange 1**
Construction	**Steel frame and plastic composite**
Habitat modules	**5m diameter, 18.8m long, 20cm walls (8 modules)**
Rotation	**40m diameter, 24 seconds per rotation, 16.7% gravity**
Population	**240 at peak, prior to ES *Shoemaker* departure**
Mass	**440 tonnes all up**

3.4 Terpsichore Station (2051)

This is a permanent station from *The Dark Colony*, orbiting a few kilometres above the surface of 81 Terpsichore, a 125km asteroid in the Main Belt between Mars and Jupiter. It rotates very slowly (30 minutes per rotation), providing less than 1% of the gravity of the Earth. Floating nearby is a non-rotating utility vessel called 'the hold' (see page 38), where boat repairs and other heavy work are carried out in free fall conditions. There is also a station on the surface of the asteroid (page 57).

The reason the gravity is kept so low is to acclimatise the colonists to the low level on the surface. This depends on my assumption that people can live reasonably healthy lives in very low gravity. There is no real science behind this, as no experiments have ever been carried out exposing people (or even animals) to fractional gravity for long periods. I considered going back to this story and making the station rotate faster, so that colonists spend some of their time in gravity, but this would mean constant movement of people between the station and the surface, messing up my plot. Sometimes in fiction, you let the plot dictate the science, but I try to keep it to a minimum.

There is no girder connecting the two halves of the station, so the process of spinning it up requires a lot of concentration and skill. It has thrusters on both halves, and station-keeping burns are carried out from time to time, with tugs assisting where needed.

Like all habitable spaces, it is built from relatively small modules, so that any loss of air (or a fire) is limited to one portion of the station. Airtight doors divide the sections, with purely mechanical closing. In case of emergency, the electric latches holding them open are released. A person might be injured getting in the way of a closing door, especially if the air pressure on one side is lower.

The modules on the outside of the station are greenhouses, exposed to light and ultra-violet from the sun, with thinner walls of translucent plastic. This makes them less safe than interior habitats, and they must be evacuated when a solar storm is expected. The colonists grow fresh vegetables there, but there is not enough growing area for self-sufficiency, so they import grain and high-protein pulses from other asteroids and from Phobos.

Eagle-eyed readers may note that the plans above show a small hall, reaching only two stories high, while other illustrations show it as three stories, as in the novel. The plans represent an earlier iteration of the design; maybe it was that

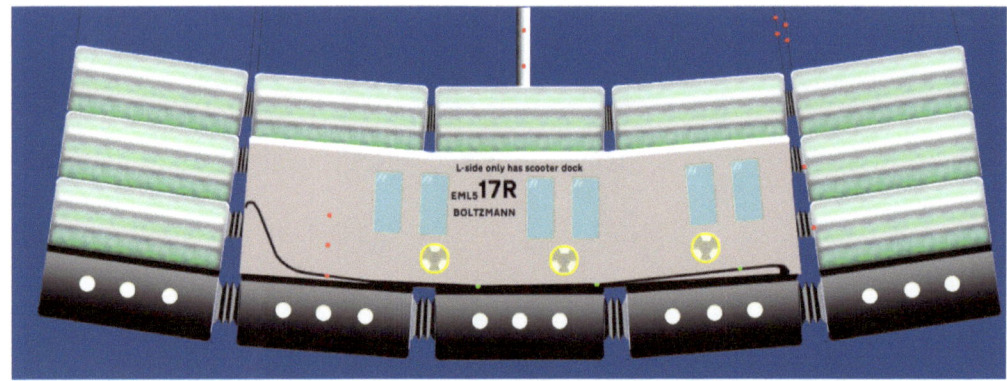

way when Lisa was a little girl. Frankly, I added the extra floor so she could vault off the balcony.

A watch is kept on the station using radar, for any incoming rocks or debris from earlier operations. Depending on the velocity, an incoming rock may also be headed off by a tug, flying out to match velocities, like a barge capture.

The view below shows a section of that early station; the ladders and hatches are in the interior, not on the outside. In early iterations I tended to write on the outside of these stations. I later realised that was foolish; millions of kilometres from their

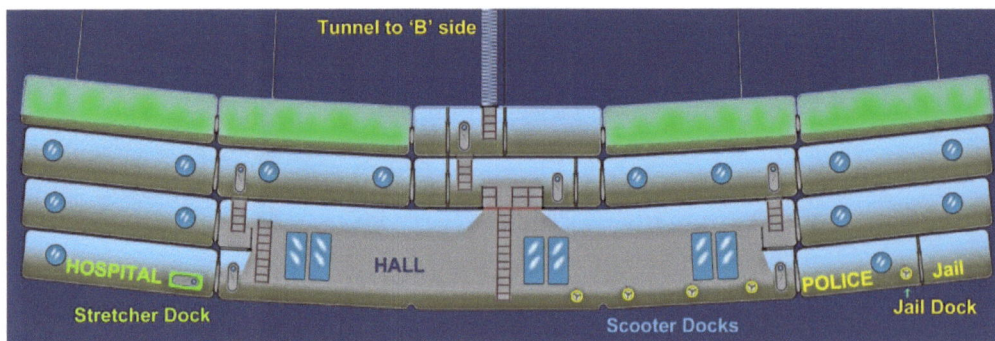

neighbours, these stations are literally only seen by the people who live in them. All identification should be inside the hall, where it will show up on video.

The hall shown on the next page is in daytime configuration, with half the area available for adults and half set up for the school. Children make up a small proportion of the population on Belt colonies; there are long periods when radiation levels are too high for pregnancies, and the colony cannot accommodate rapid growth in population.

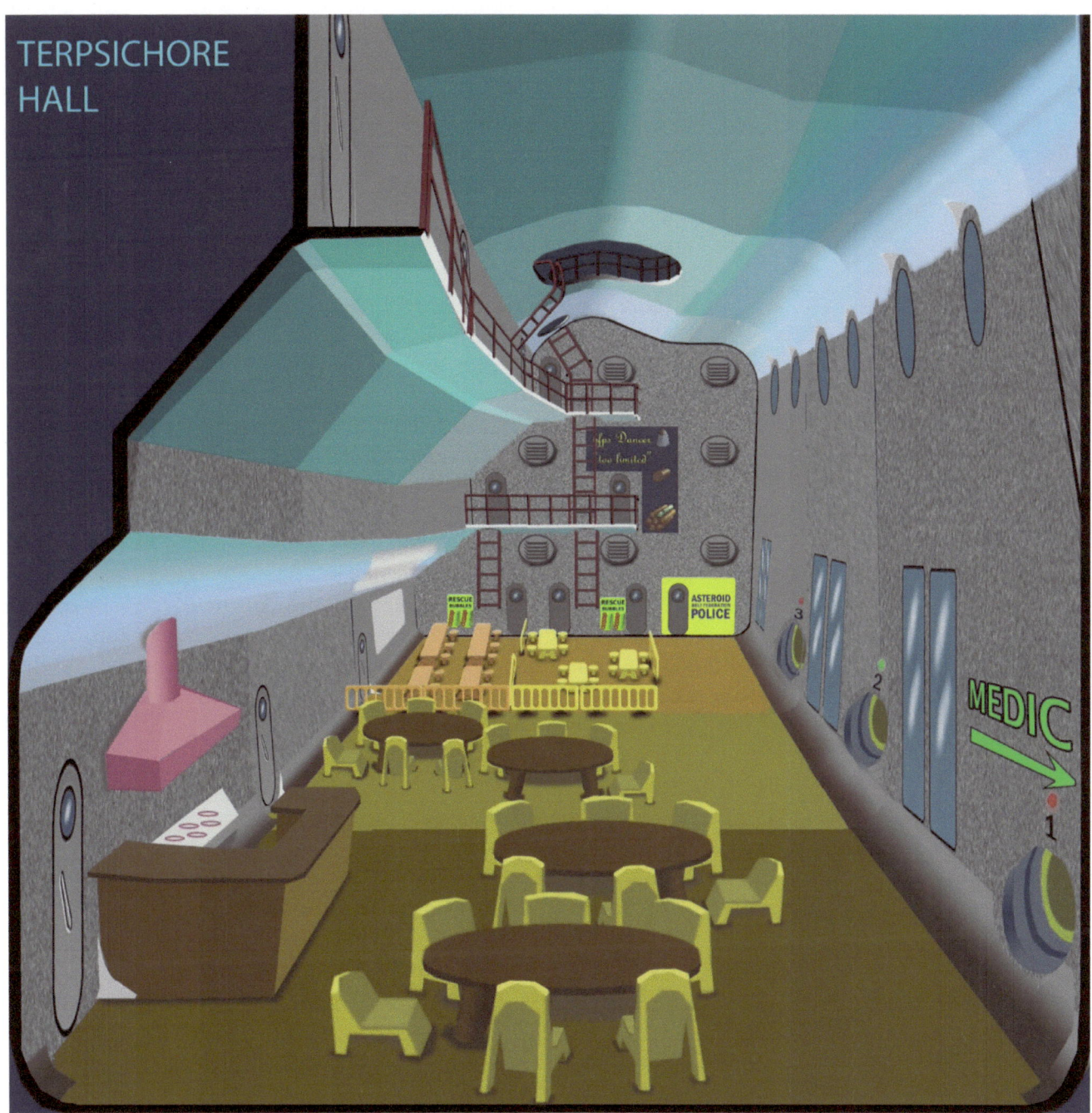

As well as a spinning motion to provide gravity, the station precesses, taking 24 hours for the plane of rotation to turn. This provides night and day to each of the banks of greenhouses; one side has a day which corresponds to Earth time, while the other is light during the time most colonists are sleeping. The hall has big windows, all on one side to provide a sense of day and night to the colonists, while interior temperature and lighting are also adjusted. The view in the windows rotates every half hour, so sunlight patterns would seem very strange to us; of course the colonists are entirely used to it.

Like the space stations of classic science fiction, the floor of the station curves upwards, without forming a complete ring. However, gravity feels vertical everywhere; you always stand at ninety degrees to the floor.

Note the hole in the ceiling: that is where you emerge from the tunnel to side B. The very first scene in my first book, *The Dark Colony*, begins in this tunnel. Climbing up to it, you feel a gradual fading of 'gravity' as you approach the hub. Passing the hub, you are weightless. Near the far end, you need to turn feet first or your head will be downward.

The central element of the station is a nuclear fission reactor (I am assuming fusion will remain ten years in the future as it always has). This generates electricity and splits water and other chemicals, providing usable chemical fuel and oxygen for thrusters, boat rockets, and heating.

Vessel	**Terpsichore Station**
Commissioned	**2047, Hamburg (the asteroid formerly known as Hamburga)**
Deployed	**Hamburg, TC46916[5], Terpsichore**
Construction	**Carbon-fibre cables and plastic walls**
Habitat modules	**4m diameter, 24m long, 12cm walls (140 modules)**
Rotation	**300m diameter, 720 seconds per rotation, 0.003 g**
Population	**220 in transit, 108 once surface colony established**
Mass	**6200 tonnes**

---

[5] A "transit asteroid": a small asteroid which passes close to Hamburg and later to Terpsichore. See page 62. 46916 is a real asteroid, with a suitable orbit.

3.5 Feronia Station (2053)

The orbiting colony at 72 Feronia, another main-belt asteroid, is very similar to that at 81 Terpsichore, and has only a minor role in the stories. The Feronia Colony (page 57) is where all the trouble lies, on the surface of the asteroid.

A unique feature of Feronia station is the full-height mural of two trees on the wall of the hall. The colonists brought seedlings of real trees to eventually grow in their place.

The station was built at **High Mars**, a station in a very wide orbit around Mars. This is where passengers and ships arrive from Cycler flotillas like Arkady (page 33) and where they set off for the asteroid belt. On the edge of the planet's gravity well, High Mars is the most economical entry to the Mars System.

Vessel	**Feronia Station**
Commissioned	**2048, High Mars**
Deployed	**High Mars, TC117984[6], 81 Terpsichore**
Construction	**Steel frame and plastic composite**
Habitat modules	**4m diameter, 24m long, 12cm walls (160 modules)**
Rotation	**400m diameter, 360 seconds per rotation, .012 g**
Population	**320 in transit, 158 once surface colony established**
Mass	**7100 tonnes**

---

[6] This transit asteroid passes close to Mars and later to 72 Feronia. See page 65.

Vessels and Stations of Earthspace and the Belt — *Richard Penn*

3.6 Hazel Stone Station (2055)

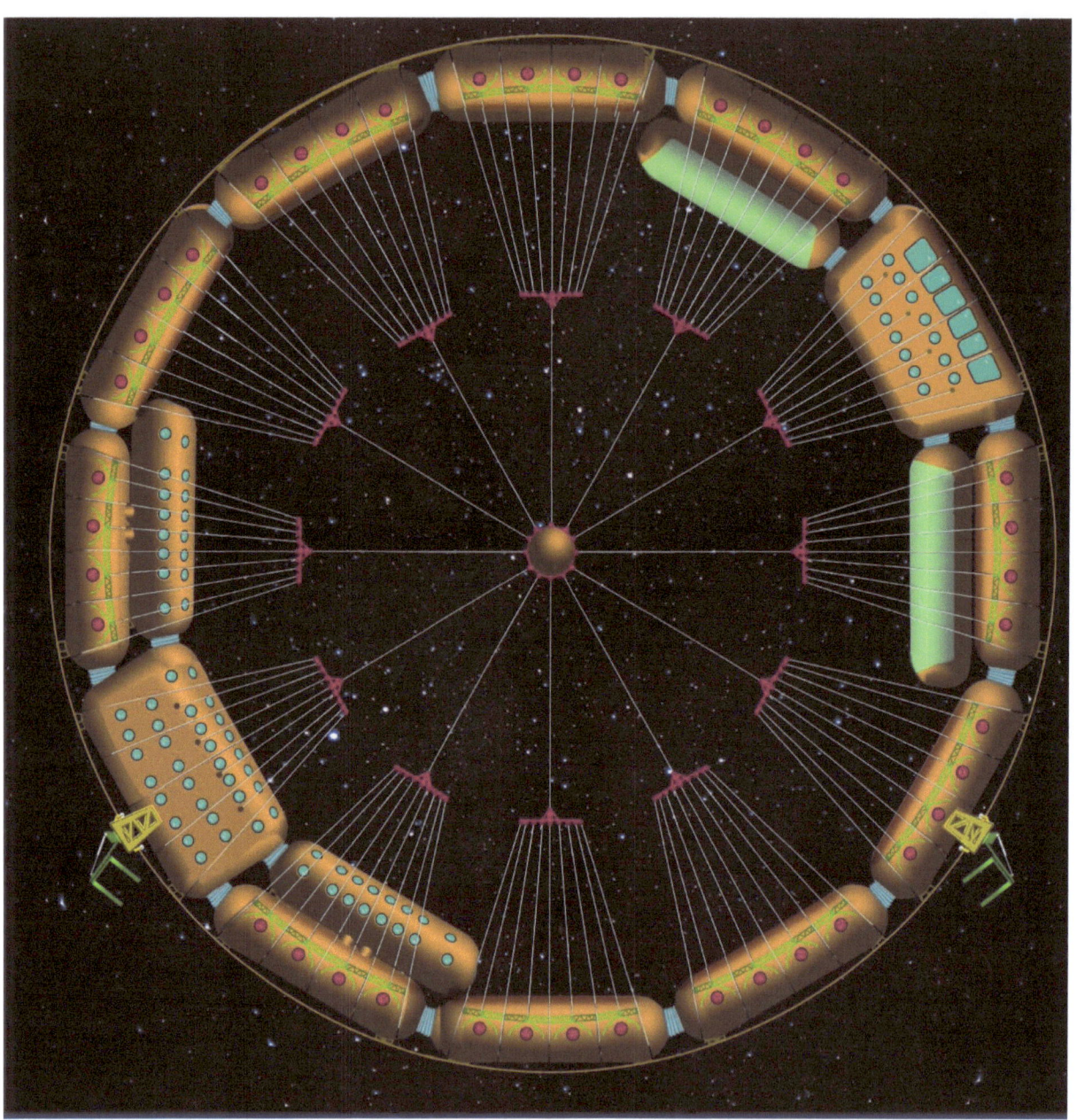

Vessels and Stations of Earthspace and the Belt *Richard Penn*

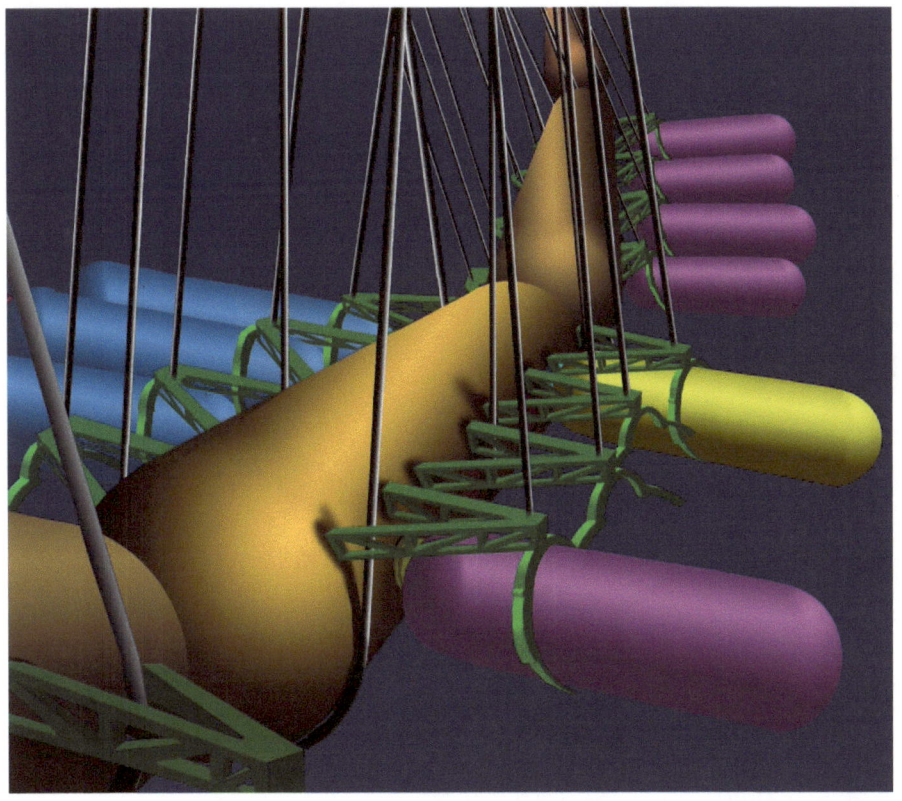

This is the most complex station, and the largest, to appear in the novels. It is one of the elements of the Arkady Flotilla, heading from Earth to Mars in 2053, and Lisa's next destination after Feronia. If the novel *Traders of Arkady* is ever completed, that is where it will appear.

The main station is a linked ring of habitats, like *Pharos* but larger. The big innovation is that it acts as a dock for up to 80 other habitats, either ships' buses (see page 20) or purpose-built live-aboard barges. It has a complex catcher mechanism running on rails on the outside, to place habitats on the dock. There is some common space aboard the station, but most visitors live aboard their boats, trading with passers-by in booths on the dock.

To dock with the station, the pilot positions the boat outside the ring for the crane to catch as it rotates by. The crane then motors around the ring to the target slot, swinging the load up and in to the clamp.

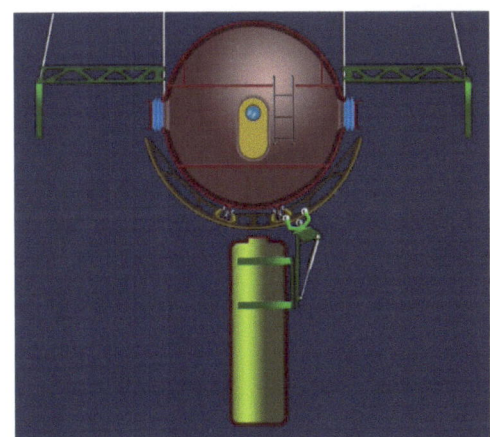
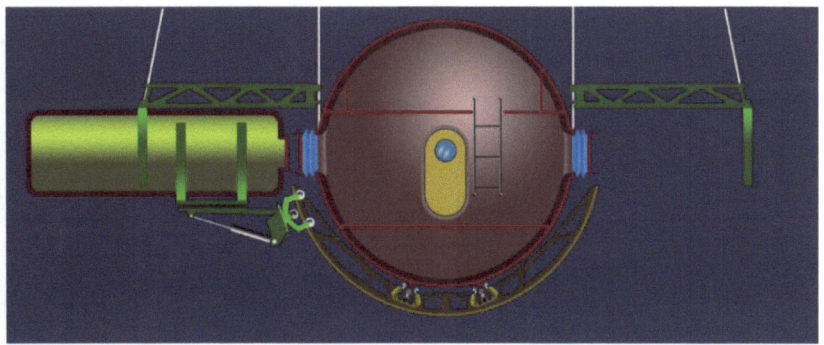

The interior looks something like a European city station: a blend of luxury shops and utilitarian offices, linked in this case by a circular mall.

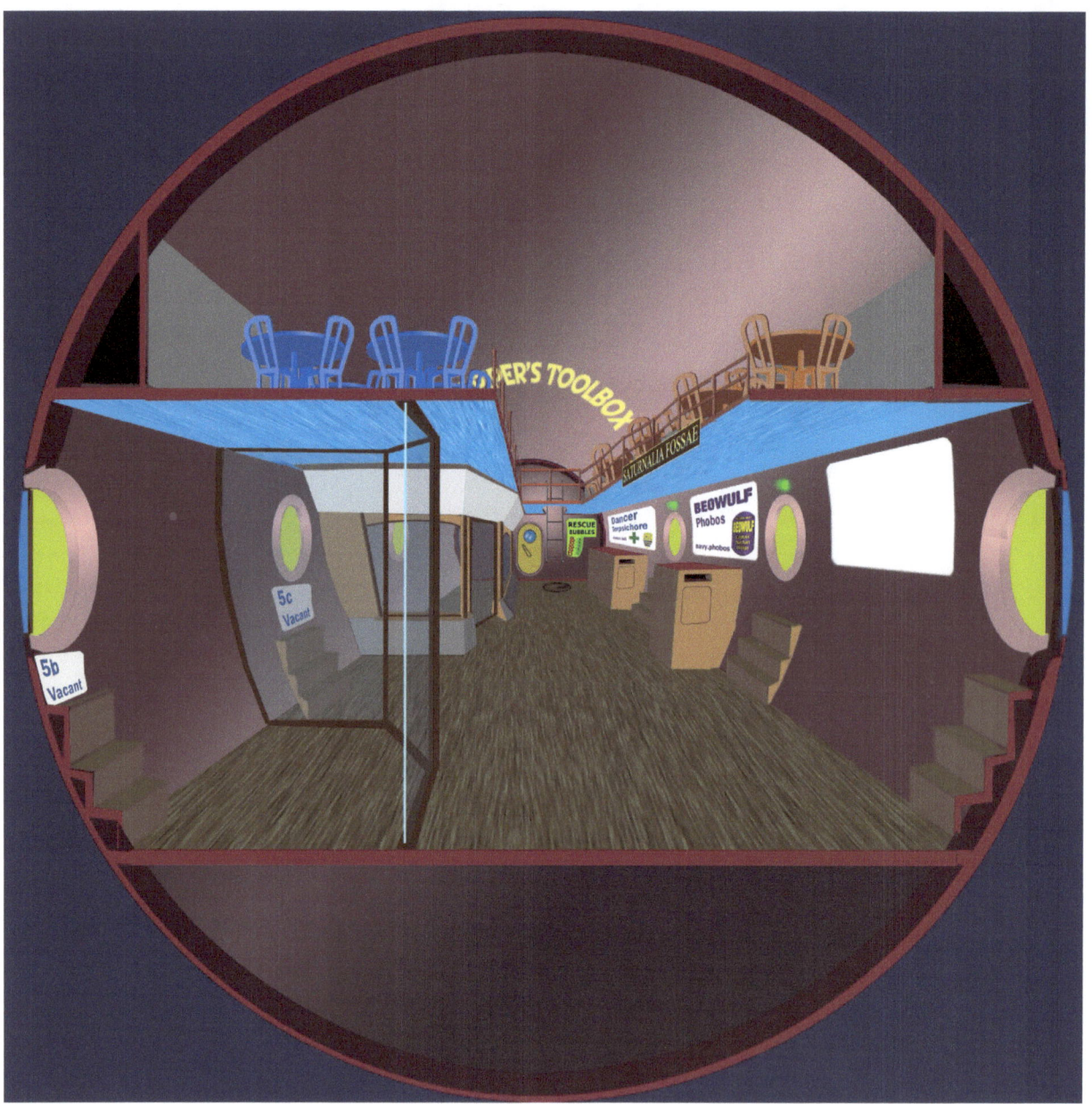

This part of the story is as-yet unwritten, but you can bet Jique is going to find kindred spirits in that basement area, and other crew will discover the joys of shopping for the first time in their lives.

The nature of the Mars Cycler orbit (see page 78) means that the flotilla as a whole is actively transporting people from Earth to Mars for a nine-month period. It then spends over five years between the orbits of the two planets but on the other side of the Sun, never meeting either.

By then it has lined itself up for a Mars-Earth transfer. Far fewer people want to make this trip, so at those times the flotilla mostly consists of barges returning exports from Mars. There follows a six-year repositioning spiral, when the crowds arrive for the next Earth-Mars transfer and the cycle repeats. A limited number of people stay aboard permanently, either to direct and maintain the station, or just because they find it a neat place to live if they can pay their way. Naturally, rents are low during the idle years.

Vessel	**Hazel Stone**
Commissioned	**2042 *Pharos*, at EML-1**
Deployed	**Arkady Flotilla, in a Cycler orbit between Earth and Mars**
Construction	**Steel frame and plastic composite**
Station modules	**5m diameter, 24m long, 18cm walls (18 modules)**
Docked boats	**~ 3.4m diameter, 15m long, 16cm walls (80 modules)**
Rotation	**38m diameter, 30 seconds per rotation, 0.17 g (Moon-like)**
Population	**450 in active transit, ~30 on recovery orbits**
Mass	**1540 tonnes empty, 4210 fully loaded**

People who know orbits will recognise that the orbit described here does not match to any established Mars Cycler scheme, though it is similar to the one invented by Buzz Aldrin, the Apollo astronaut. This is because I envisage the flotilla being under continuous thrust in idle times, using nuclear-powered ion rockets. I haven't been able to find published data on cycler orbits using continuous-thrust spirals, so I made up one that works for the stories. I do time the Earth-Mars departures to Hohmann-like economy windows between the real planets.

4 Utility Structures

4.1 Copenhagen Spacetug

This private space station, set up by the engineering co-op in *Spacetug Copenhagen*, is little more than a series of cargo modules and tiny habitats, assembled by hand and bolted together, as described in the first *Steps to Space* novella. The modules are delivered by individual launchers, to orbit passively near the luxury hotel. Once assembled, it becomes the work area, fuel dump and junkyard for the satellite repair business. The engineers live in the former hotel, which provides gravity and a more comfortable environment. This establishes the pattern for later colonies, with a spinning station surrounded by a loose "cloud" of workshops and cargo containers. The expandable module, which can be opened to space at one end or closed off and filled with air, is a work area for assembling boats, known as the hold.

At this stage, solar power is used, with cells reclaimed from disused satellites. Later, this was replaced by nuclear fuel, as solar cells have to be imported from Earth.

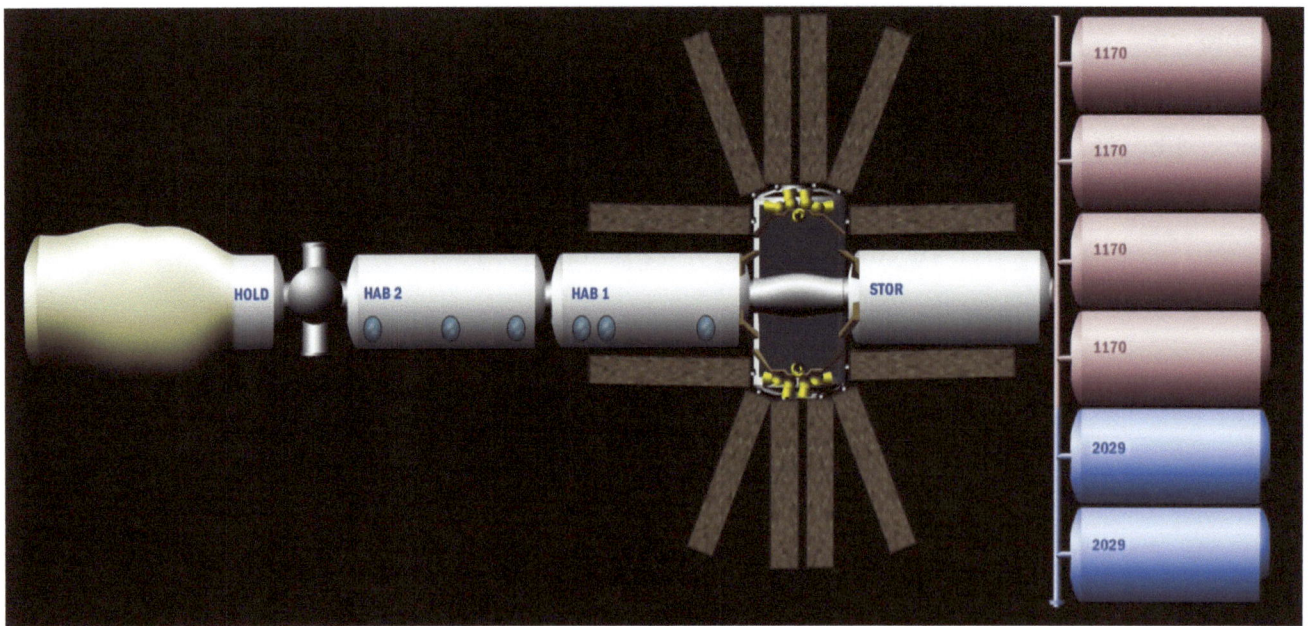

4.2 Terpsichore Hold

The hold, in the later asteroid colonies, is where the engineers spend their working days. All work on spaceboats is easier in zero gravity than in a spinning station, and the hold has a boat-sized airlock at each end, so the work can be done in air rather

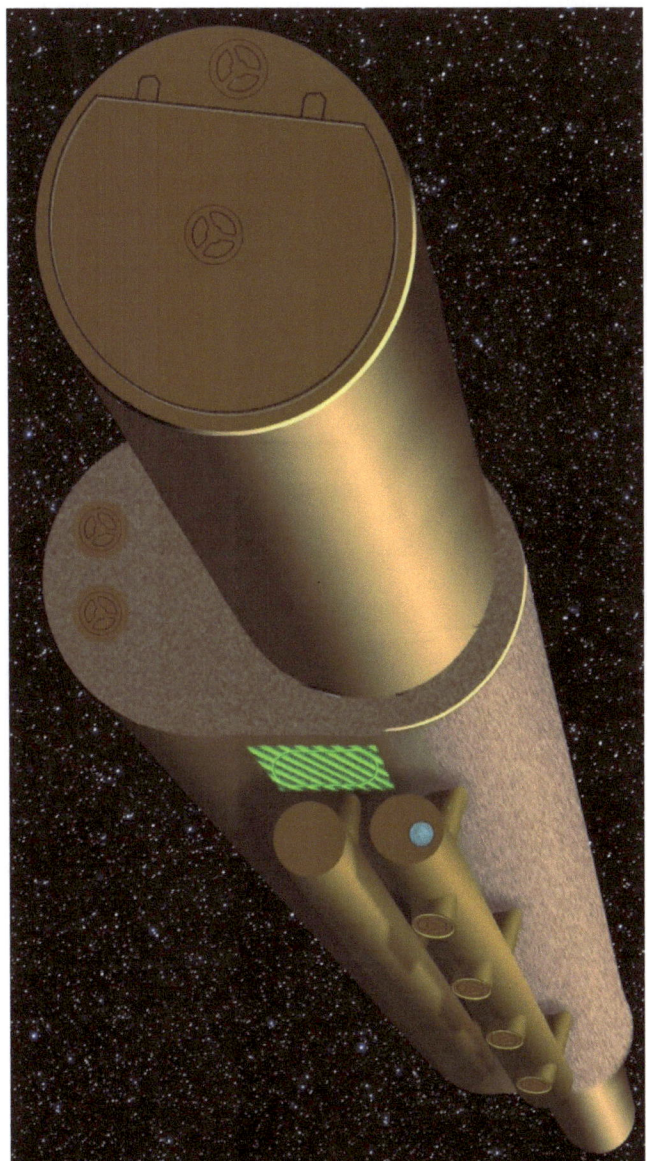

than vacuum.

Once inside, boats are moved by hand, using chains and ropes the way a similar-sized boat might be handled at a boatyard on Earth. The term 'lockhand,' which I borrowed from the English canal network, refers to a semi-skilled worker, usually young, who moves things about in the hold. Lockhands work as apprentice mechanics, on the pathway to becoming a hands-on engineer. I do not believe in the idea of an engineer as a pure desk worker who never gets their hands dirty.

4.3 The Cloud at Pharos

Many of the people who go into space are independent-minded, libertarians even, unlikely to be comfortable in the crowded, civic-minded confines of a station like *Pharos*. These become the "rockhounds" of sci-fi tradition, who aim to prospect and mine small asteroids or remote parts of the Moon, meeting in disreputable bars at various ports.

In *Mutiny near Earth*, the cloud is built as a holding area for cargo, like a container port on Earth, but rockhounds and other informal travellers move in, turning it into a sort of caravan park as well.

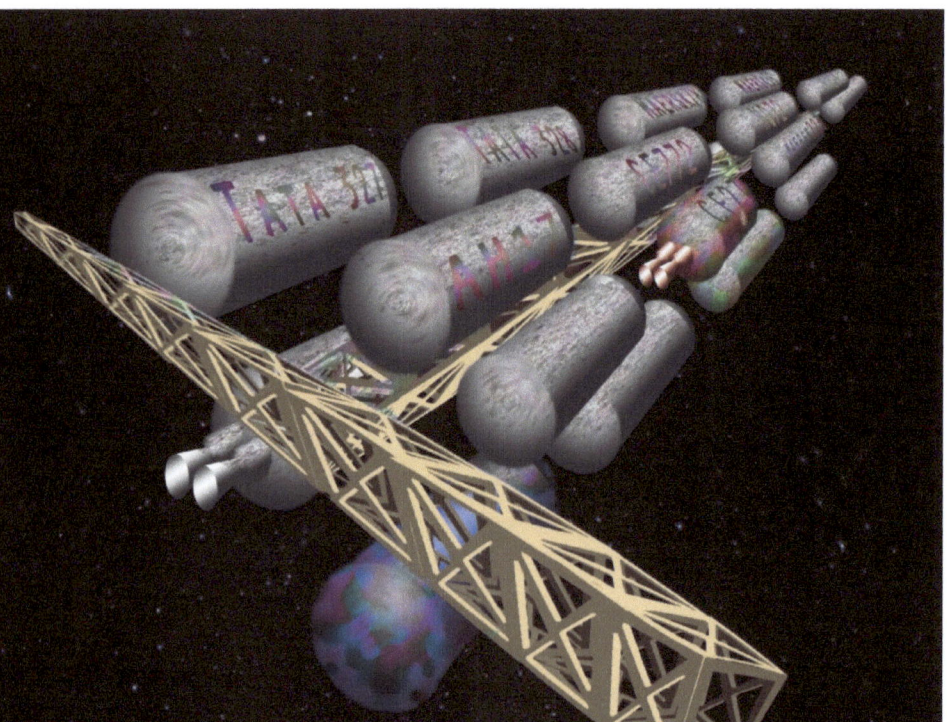

Like anything at EML-1, the cloud needs an occasional station-keeping boost to stop it drifting away and failing towards the Moon or Earth. This service is provided by *Pharos* station using tugs, twice in each 13-day orbit. Otherwise, the denizens of the cloud are left to themselves, trading with each other and with the station.

5 Ships

Throughout science fiction, the word spaceship has been in use since the beginning, but **spaceboat** has never even made it into the dictionary. I draw a line between ships and boats roughly where it is drawn for seagoing vessels; a ship is permanently crewed and goes on long journeys, while a boat is used for a day or two, and stays around a station or a ship.

5.1 BFPS Dancer (2051)

This is the chemical-powered warship (more of a boat, really) cobbled together in *The Dark Colony* from a tug and some cargo containers. She is meant for a journey to an asteroid in almost the same orbit as 81 Terpsichore, taking only a few weeks.

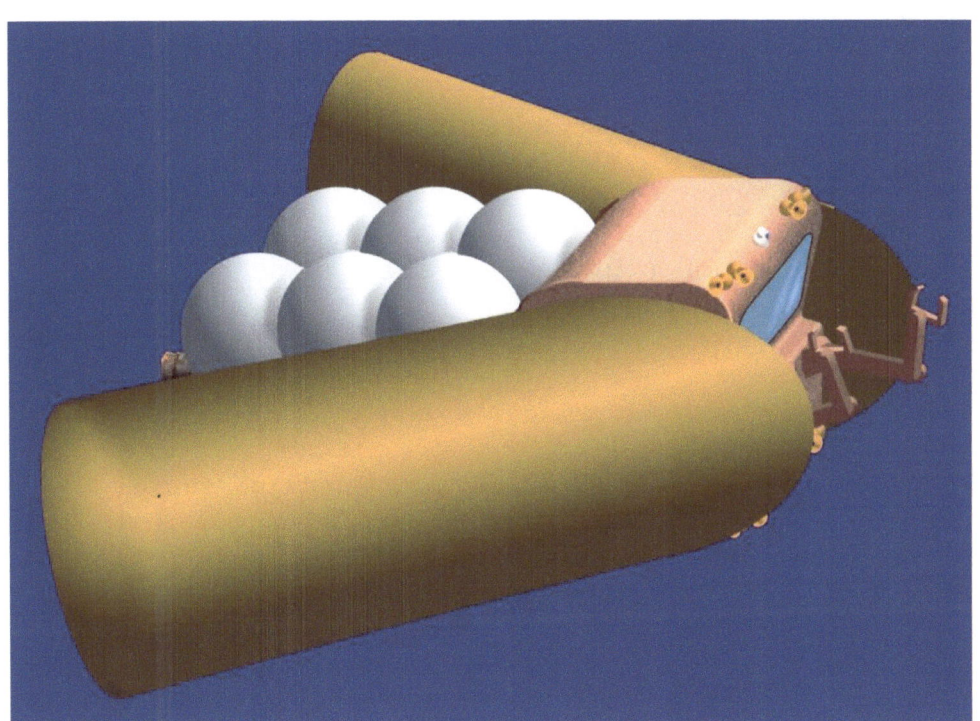

The rockets and all other systems run on chemical power, but she carries two Radio-isotope Thermal Generators (RTG) so she can be refuelled from supplies of water ice found along the way. This allows her to have relatively small fuel tanks (good for about 1 km/s of delta V), and then extend her mission indefinitely as long as she can find ice.

This picture is an early concept sketch; the final design was longer and used more tanks.

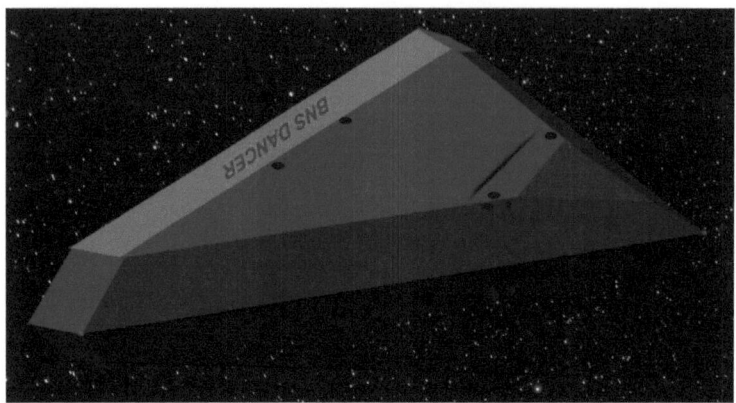

She has a stealth shield to make her less visible to radar, made of thin metal.

Internally, she is quite cramped; the cabins in a railway sleeper car on Earth would be roomier. She carries two boats; the Captain's Gig (an armed scooter, see page 11) and The Blob, a flimsy plastic vessel that prioritises stealth over safety. Importantly, these boats are docked to the main habitat of the ship, so they can be accessed without spacesuits if needed.

Vessel	**Belt Federation Police Ship *Dancer***
Commissioned	**2051 Terpsichore**
Deployed	**The local group of small asteroids, co-orbital with Terpsichore**
Construction	**Nickel-iron shroud, plastic habitats, and carbon-composite tanks**
Habitat	**Same as tug, plus a 9m x 3m habitat cut slantwise**
Docked boats	**Armed scooter, plus the Blob, a 500kg all-plastic boat**
Engine performance	**Same as tug**
Propellant	**Chemical fuel and oxidant. 24 tanks, 1413kg empty, 10 tonnes**
Effective range	**Total delta V 1.21km/s (in the story, she is launched by tugs)**
Crew	**6**
Mass	**28 tonnes empty (including boats), 38 fuelled**

The total delta V shown is quite low; with the extra habitat, the shroud and the fuel machines, this is not a nippy vehicle. Her mission is to asteroids in Terpsichore's Local Group, kilometre-sized asteroids in the same orbit around the Sun but leading or trailing the asteroid itself. These are fictitious, though I would argue that such things might exist, based on the trails of meteors that precede and follow comets in Earth-intersecting orbits, made up of debris from ancient collisions. If they do exist, they would be too small for us to have detected them, so like gods and unicorns I can say that science has not proven them unreal.

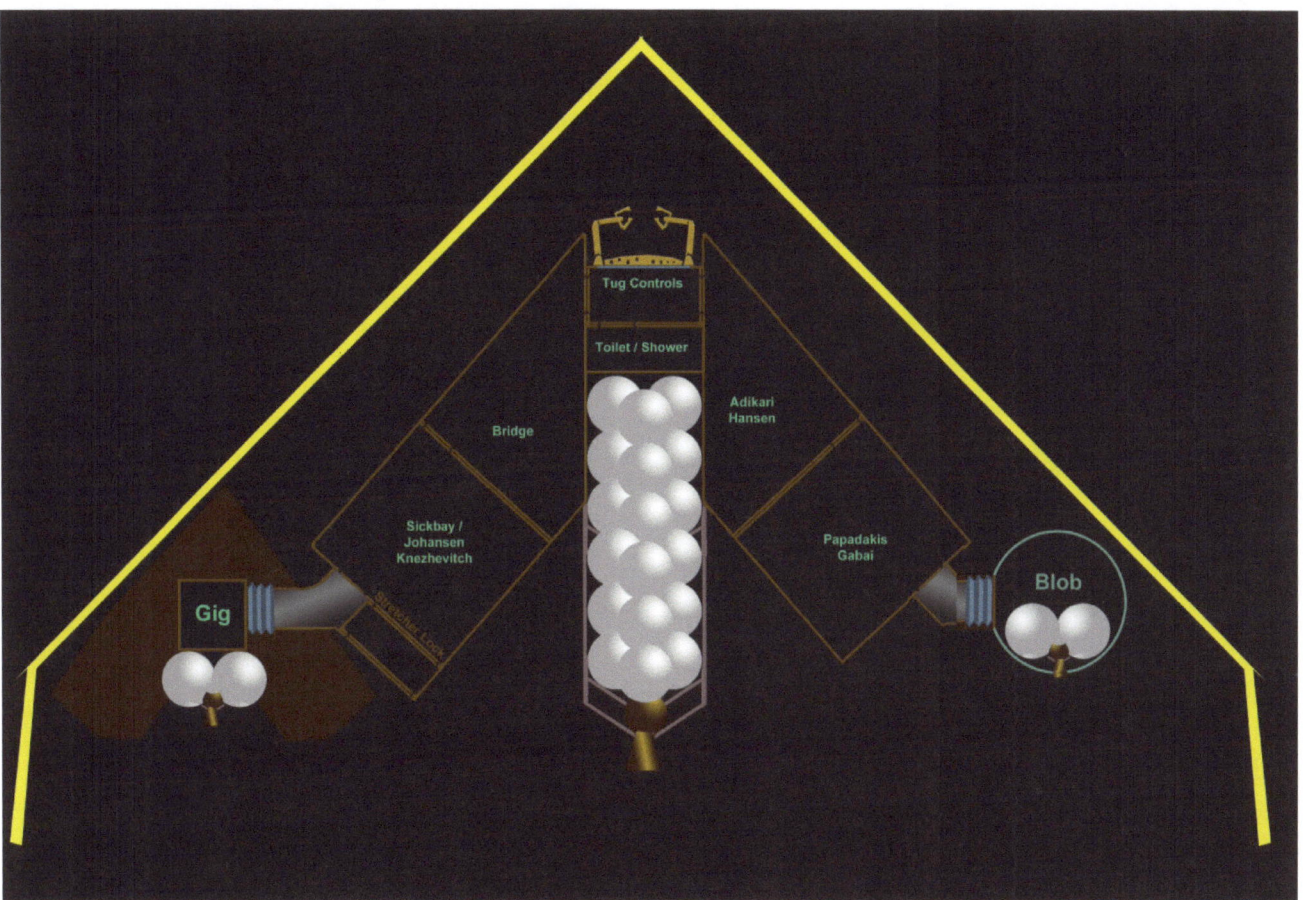

The number of tanks here is more accurate than the previous picture; she is supposed to carry 24. Not shown here are the ice machine and the fuel machine; RTG-powered devices for extracting rocket fuel from the ice, that reside behind the habitats on each side.

5.2 BNS Dancer (2053)

The second ship named *Dancer*[7] was designed by Tommy Hansen and the crew for the years-long trip from 81 Terpsichore to 72 Feronia, and on to Mars. She has a nuclear-powered ion-accelerator engine, built to blast for a few days, using water as propellant. For the rest of the time, she coasts, separated out into three components and spun up for gravity. When the main drive is needed, she has to be spun down and stacked, as shown below. She cannot use her main drive while spinning, and the gravity is quite low, about 2% of Earth gravity.

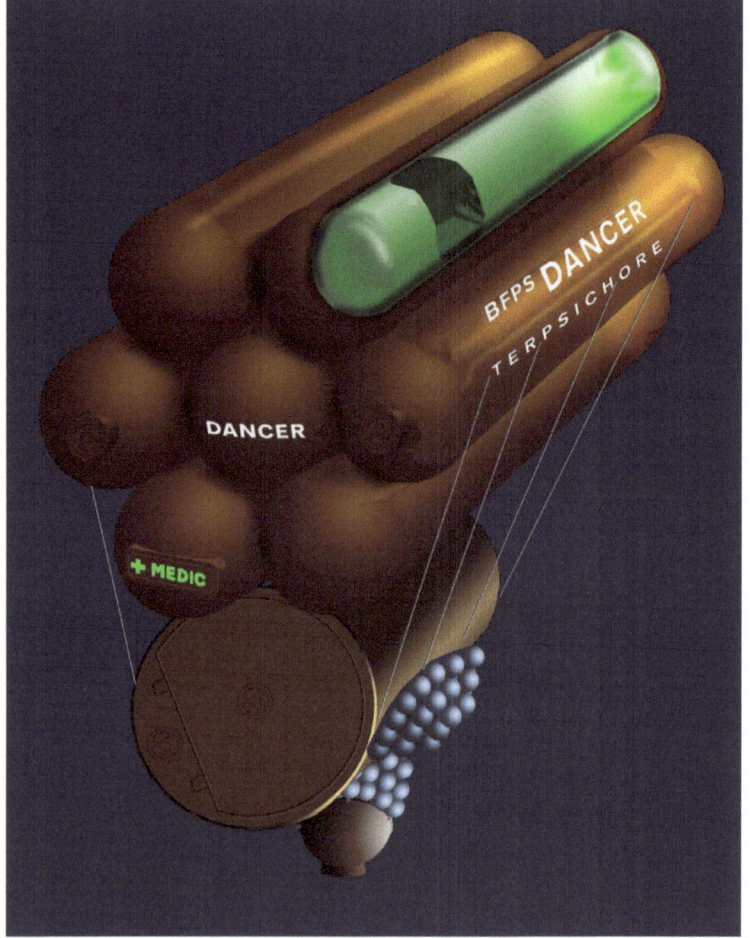

---

[7] Terpsichore, in classics, is the spirit of the joy of dance, hence the name.

Internally, because of the tight stacking of the habitat cans, everything is very close together. Most of the compartments can be reached from the main hall by hatches in the floor or overhead. In Earthling terms, it would be like living on a seven-car train where the carriages are tightly stacked instead of strung out in a line.

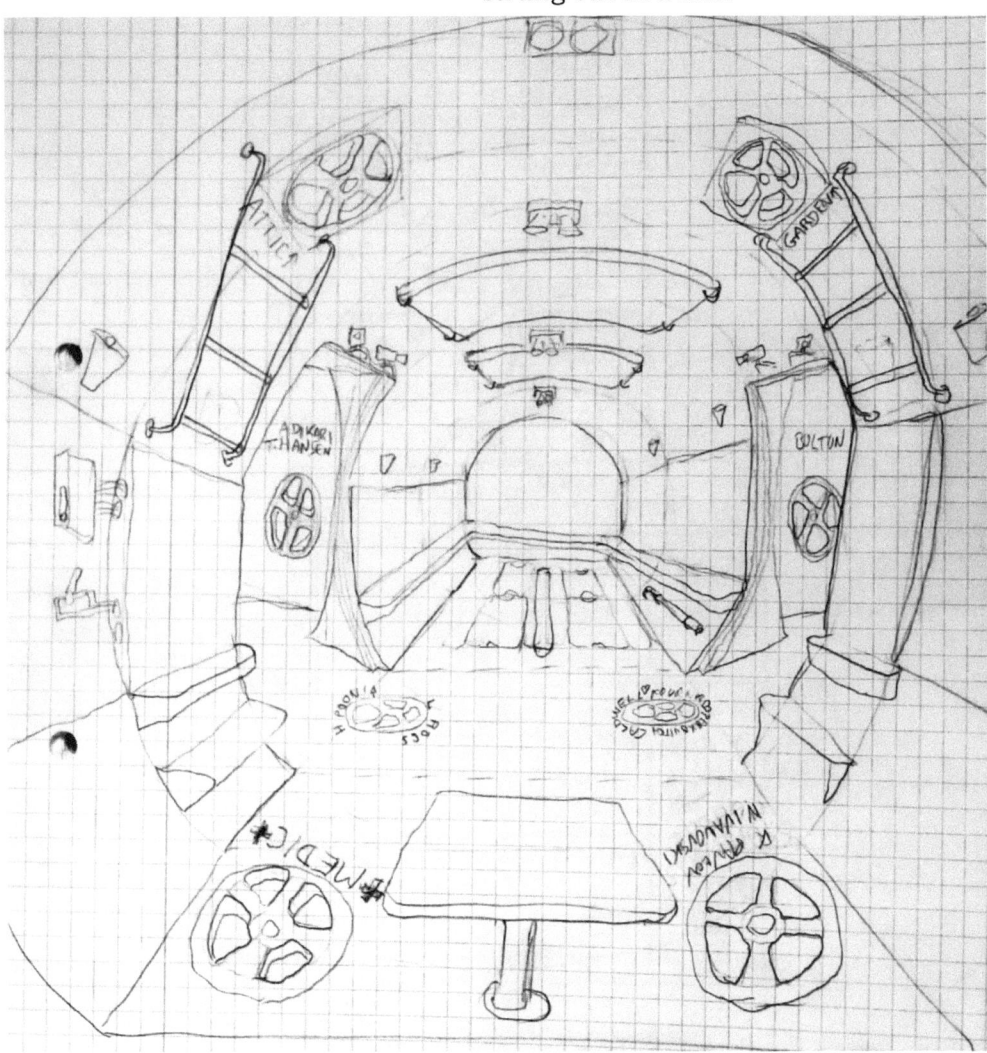

Most of the time, the ship is coasting; the duties of the watch involve little more than monitoring displays. The bridge, located in the main hall, is simply a semi-circular bench flanked by displays.

One of the upper cans has a translucent exterior for growing plants, and a clear glass window to see the stars. This mostly fills a psychological need; they carry stored food for everyday needs, though the fresh herbs and fruit are a welcome addition.

Vessel	**Belt Navy Ship *Dancer***
Commissioned	**2053, at 81 Terpsichore**
Deployed	**Transfer to 72 Feronia and on to Mars**
Engines	**Nuclear fission / ion accelerator, 21 t with shielding Exhaust 13.6 km/s**
Propellant	**Water (carried as ice) 200 tanks, 18 t empty, 340 t capacity**
Hold	**12m x 4m diameter, 10cm plastic walls. 12 tonnes**
Habitat	**7 x 12m x 2.8m diameter, 12cm walls, 143 tonnes**
Docked boats	**Two scooters, a tug and the Blob, 10 tonnes total**
Delta V	**Refuelled at transit asteroids, 12.9 km/s per leg**
Crew and passengers	**23, including children**
Mass	**215 tonnes empty, 557 tonnes fuelled**

Note that all the outer walls, including the habitat, are thinner than those used on the station. This saves mass, but makes the ship more vulnerable to both meteor strikes and solar storm radiation. The fuel tanks are also extremely light, barely more than plastic bags, as the fuel is frozen for storage.

5.3 ES Shoemaker (2035)

Earthspace Ship *Shoemaker* appears in *Mutiny near Earth*. She went through a number of flag changes during her commissioning, as she searched for a legal framework to suit both her financial backers and the people who travelled in her.

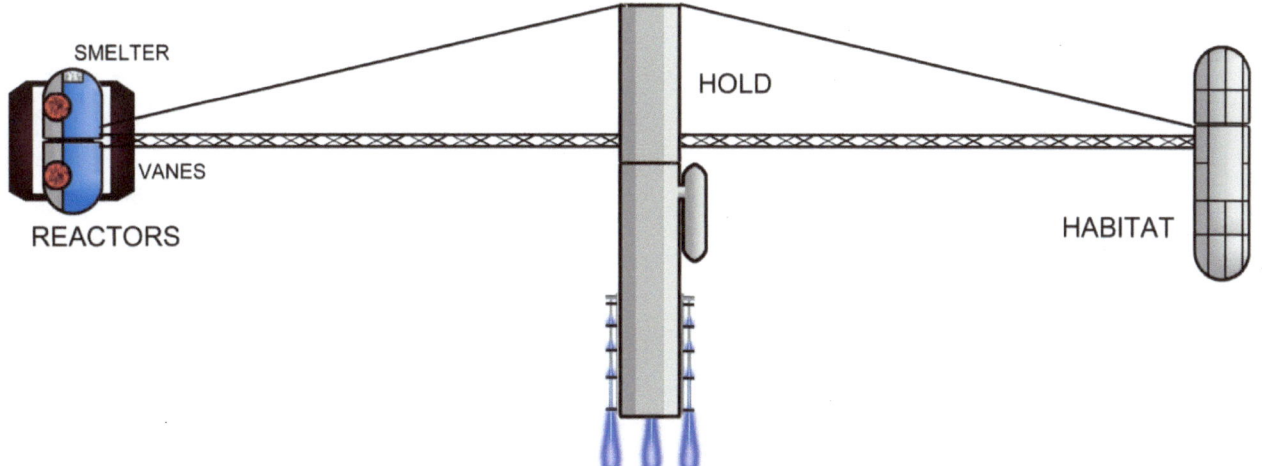

She is a nuclear-powered ion-rocket ship intended for a 2035 mining expedition to a near-Earth asteroid, the first of its kind.

The ship is divided into three hulls, each with a very different function. The central hold, with the drives mounted on it, contains all the ship's boats and mining equipment, as well as propellant tanks. Revolving around this on slender girders are the nuclear reactors at one end and the habitat at the other: a very crowded home for eighty people for more than a year.

Near-Earth Asteroids (NEA's) are actually not very near for most of the time. They get their name from the fact that their orbit is inside that of Mars, unlike the more numerous main-belt asteroids. But at any given moment, a NEA may be anywhere in that broad zone, up to 400 million kilometres away on the other side of the Sun. Some NEA's pass relatively close to the Earth (a few million kilometres) at one point in their orbit, but that approach is only once in many years. In order to get to one and bring back useful materials, the round trip must a year or more.

At the time of my story, several uncrewed missions have visited asteroids. Most of these, like nearly all robot missions, either failed completely or returned tiny amounts of material. In my story, the participants rely on human resilience and ingenuity, and aim to use that to bring back a lot, to make large sums of money and build more ambitious projects in space.

This picture shows the ship departing from *Pharos* (see page 24) for the first time, heading for a gravity sling around the moon and out to the asteroid.

Most of the crew spend all their time in the habitat, in 10% of the Earth's gravity. The ship rotates once per minute, quite fast compared to later stations, as people were unsure of the health effects of very low gravity. They adapt the ship during flight based on experience with 0.1 g, removing stairs between levels, for example, because it is easier to jump.

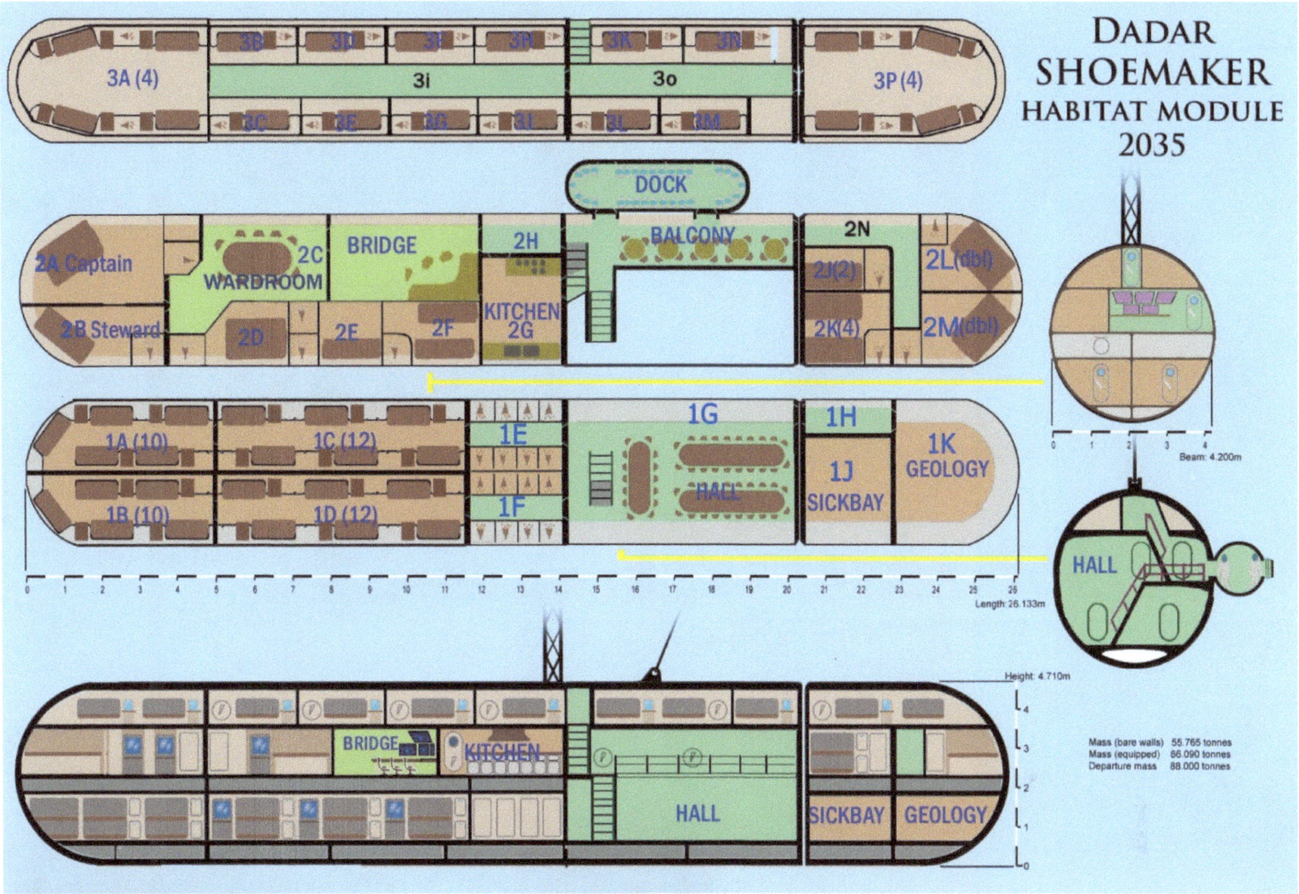

DADAR SHOEMAKER HABITAT MODULE 2035

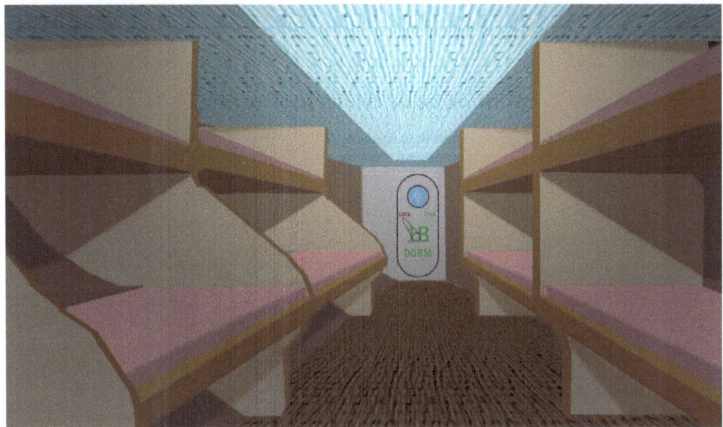

The habitat is three stories high, though the attic level is built into the curve and has little headroom. Most of the accommodation is in dormitories of up to ten people, with only the officers and professionals having private rooms. The dorms are just as cramped on military submarines today, where people manage to tolerate long missions.

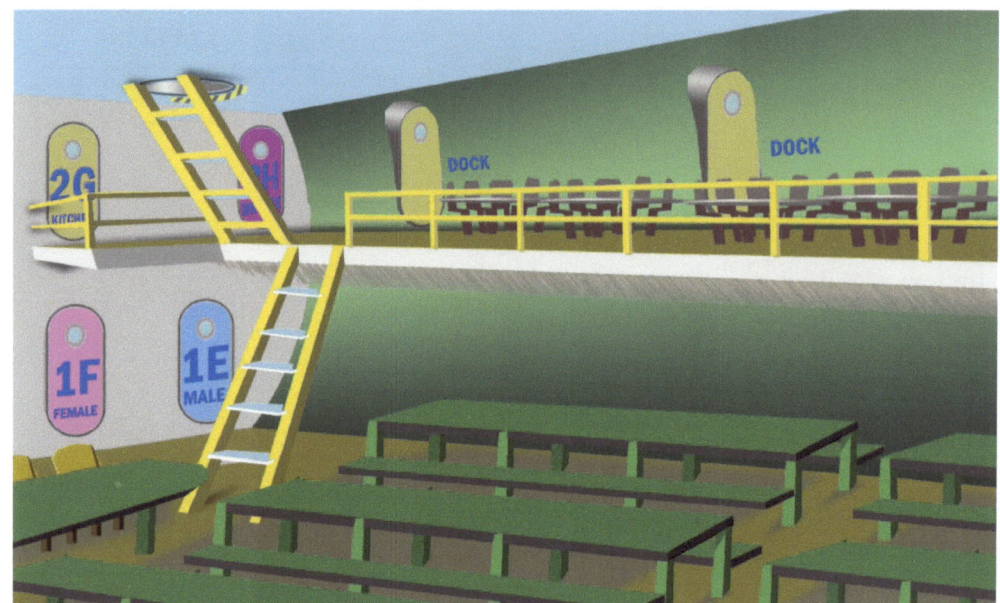

Everyone spends most of their time in the central hall, acting as canteen and social area. The exterior of the hull is 12cm solid plastic, with no windows except for a panoramic one at the front of the ship. This is not intended as a bridge (looking out the window in a spaceship is rather pointless) but as a supervisory office once they reach the asteroid.

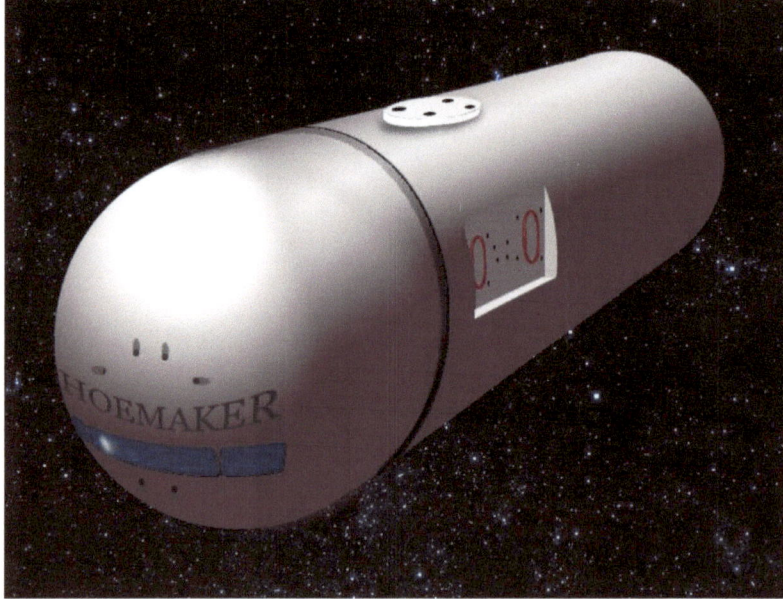

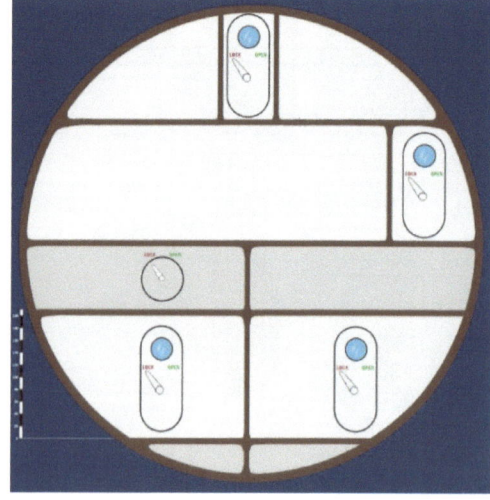

At that time, the ship will be disassembled, the rear half staying in orbit for the crew while the front of each module is parked on the surface.

Vessel	**Earthspace Ship *Shoemaker***
Commissioned	**2035, built at Tycho, on the Moon, assembled at *Pharos*, EML-1**
Deployed	**To mine a near-Earth asteroid on a 2-year mission**
Engines	**Nuclear fission, 2 reactors 26 t with shielding** **4 ion accelerators exhaust velocity 6 km/s**
Propellant	**Xenon carried in the hold, 226 t capacity, 8t tanks (refuelling with water for the return flight)**
Hold	**40m x 5m octagon, 15cm plastic walls. 55 tonnes**
Habitat	**27m x 4.7m diameter, 15cm plastic walls. 77 tonnes**
Docked boats	**Two scooters, two tugs, 12 tonnes total**
Delta V	**Refuelled at target asteroid, 4.8 km/s outward**
Crew and passengers	**80, including 5 children**
Mass	**214 tonnes empty, 476 tonnes fuelled**

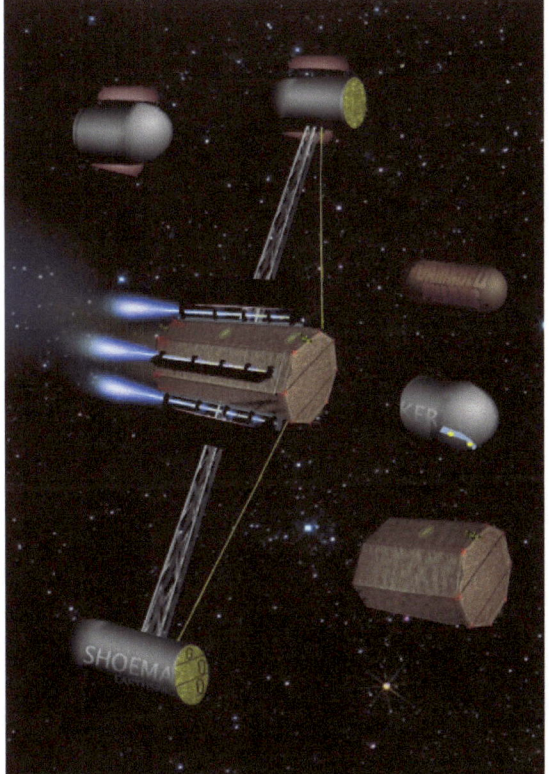

Like all ion engines, *Shoemaker*'s main drive has limited thrust, providing less than .001 g with full tanks. Leaving EML-1, she takes several days of spiralling orbits before falling towards the Moon for a gravity assist to send her on her way. She could never land on the moon, or even an asteroid more than a kilometre or so in size.

5.4 PNS Beowulf (2054)

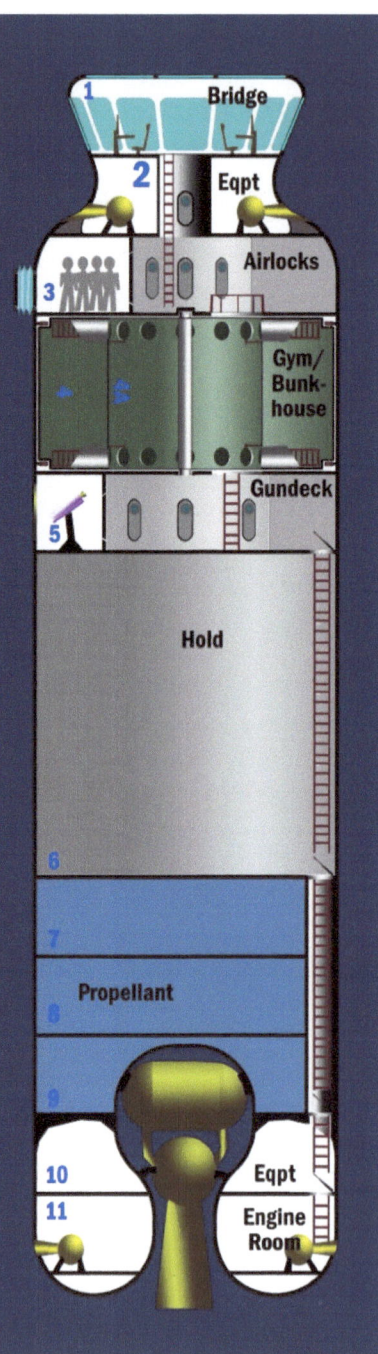

Phobos Naval Ship *Beowulf* is mentioned in passing *in The Dark Colony,* having been summoned to the aid of Terpsichore only to be called off because our heroes have dealt with the crisis themselves. She has been worked out in much more detail, for inclusion in *Traders of Arkady,* but I still have not done the maths on her performance, so things may change.

Most spaceship plans are far too delicate to be used for military ships, with fragile girders and elements strung together with cables unable to resist the simplest of weapons. This design addresses that with a unitary hull, mostly by compromising crew comfort.

Making a ship smaller means that she cannot be effectively rotated for artificial gravity, so the solution in military ships is an internal centrifuge, used as exercise area and strung with hammocks as a bunkhouse. Unless the ship is huge, that centrifuge has to have a small radius and therefore needs to rotate quickly. I assume that, with suitable training, people can tolerate this, but again there is no real experimental evidence for this.

The other thing a military ship needs is to outrun the opposition, and that means that much of the hull is given over to fuel tanks. In *Beowulf,* a moveable wall separates the tank area from the hold where boats and cargo are kept; the trade-off between the two depends on the mission.

A hard-core military ship would not have the command and control centre exposed like this – this particular ship is designed for harbour duties supervising civilian traffic, not for outright war, which has not occurred in space at the time of my story

5.5 Other Ships

Unfinished stories tend to acquire technology concepts faster than the characters can get around to visiting them; this is

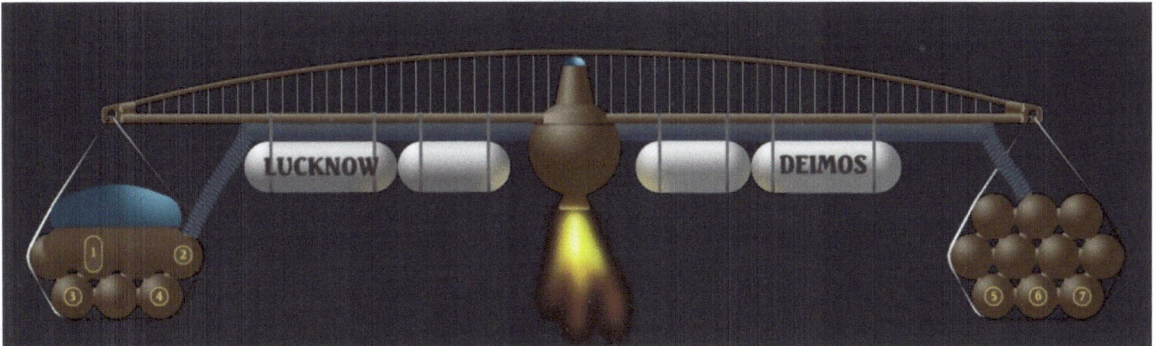

especially true of *Traders of Arkady*.

DNS *Lucknow* is passenger ship on the "coat hanger" plan, allowing quite high levels of gravity under drive, and keeping the passengers comfortable at all times. She mostly serves people from the planets or the Moon, who want to travel in space without discomfort.

Vessels and Stations of Earthspace and the Belt *Richard Penn*

At far ends of the social scale, the luxury liner ES *Clarke*...

...and the unnamed "lash-up."

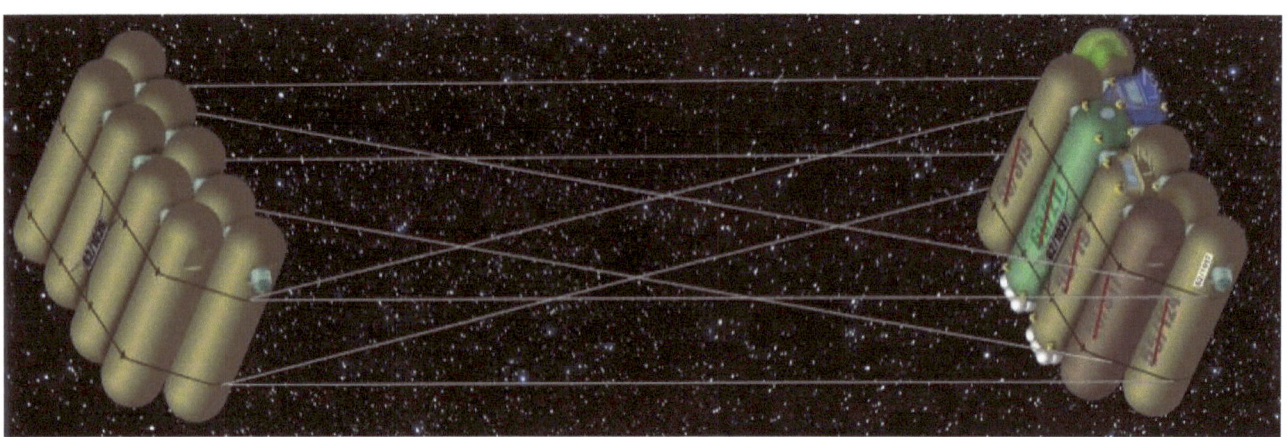

6 Surface Stations

For most space programs, getting to the surface of other planets, asteroids and moons is the whole point. Some of my main characters tend to see space travel as an end in itself, viewing a spaceship or an orbiting station as the ideal place to live. However, they recognise that is a minority view, and many of their fellow travellers can't wait to reach some kind of destination.

6.1 Moon Colonies

The first permanent surface colonies will be on the Moon. We had the technical capability to set these up forty years ago, and serious plans are being made in Russia, China and Europe. Travel times from Earth are only a few days, so you can arrange rescue or resupply missions when things go wrong. As is not true for Mars or any other destination, it would be insane to set up a colony there without first settling the Moon.

In *Caverns of Procellarum*, the only extant colony is *Qianshao*, a Chinese one[8] that is not doing well. Their 3-D printer has broken down, and their home country has suffered a revolution, cutting off their supply line.

Other colonies have been set up, with equipment delivered by uncrewed missions and assembled by robots, but none are occupied.

---

[8] All lunar background images credit NASA / Apollo 15

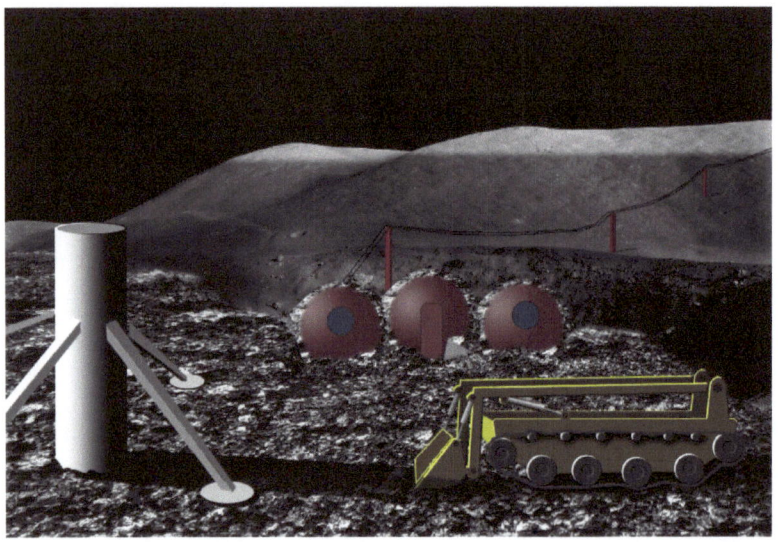

In keeping with my general preference for humans over robots, these have generally broken down; it is up to my characters to fix them. Here is a Russian one, near the south pole of the Moon.

Compared to space stations, surface colonies have a unique advantage: they can use local materials, at zero cost in "foreign currency." On the Moon, it makes sense to cover any habitat with a metre or so of soil (or regolith, to use the technical term) as protection against radiation and to even out temperature variations between day and night. Elaborate house-sized 3-d printers are not required; you can simply pile it on with a digger, or even use a bloke with a shovel.

The Moon has a number of lava tube tunnels: huge underground caverns stretching kilometres into the surface. They show the greatest potential as future human living places.

Once established, colonies on the Moon will use Earth time, initially in the time-zone of their home base, but later in a common time-zone across the whole Moon. In my stories, they fix on Greenwich mean time, or UTC. The night/day cycle on the Moon corresponds to Lunar months on the Earth (29.53 days), and is useless for humans. If you live in a tunnel there is little point in a calendar that follows outside conditions.

A useable calendar could be proposed that makes the day-night cycle exactly 28 civil days, where a "day" (like on Mars) is longer than 24 hours, but the disruption and confusion with Earth calendars would not make it worthwhile.

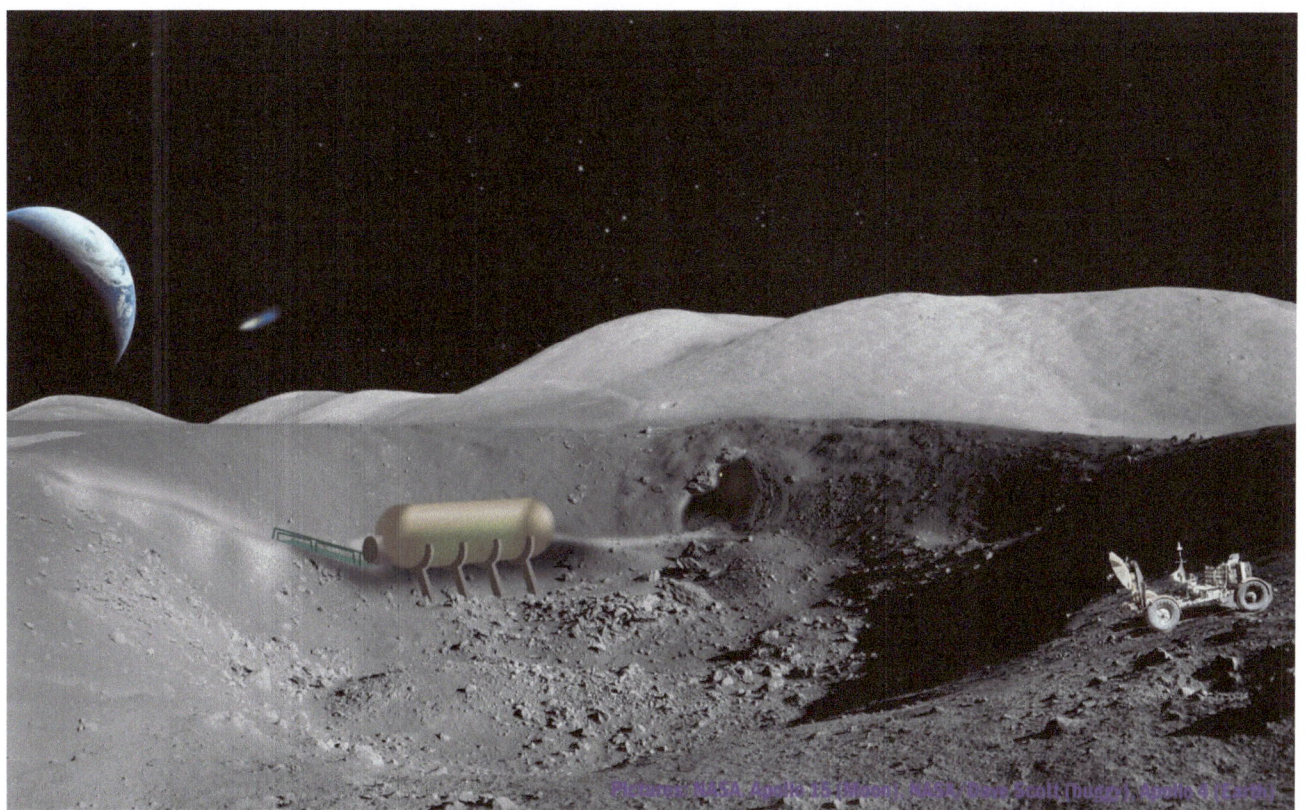

Yes, that's Apollo 15's buggy. They plan a museum one day.

6.2 Asteroid Colonies

The ultimate goal, for many of the people who will occupy space, will be a self-sufficient colony on another world, a new

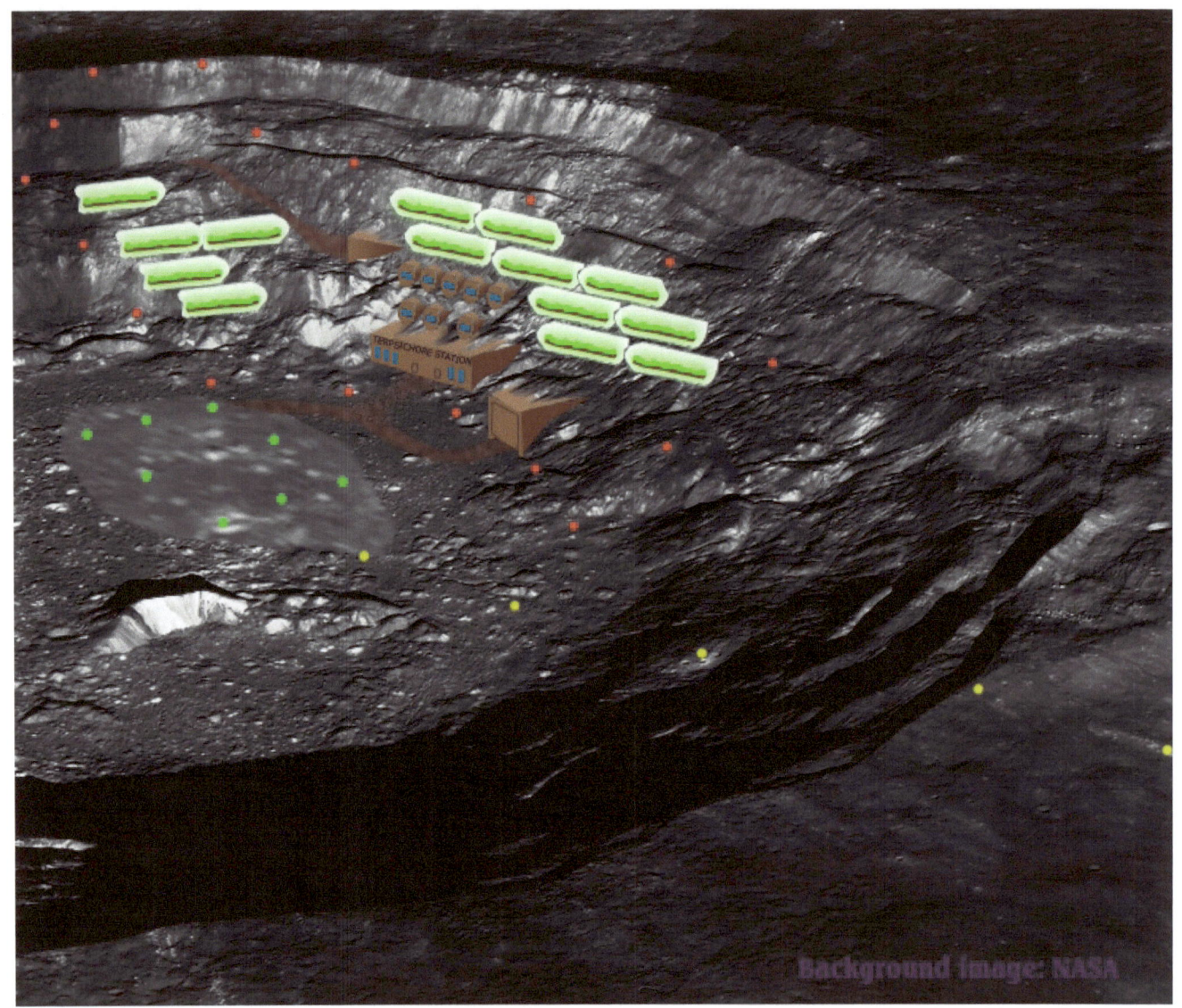

habitat for humanity. By 2050, when my first story is set, true self-sufficiency is still far away. They are growing some of their own food, but it takes a huge amount of growing area to support even a few hundred people, so they still import grain and protein foods from Mars. Also, modern technology is hugely complex. They try to keep what they have as simple as possible, and build

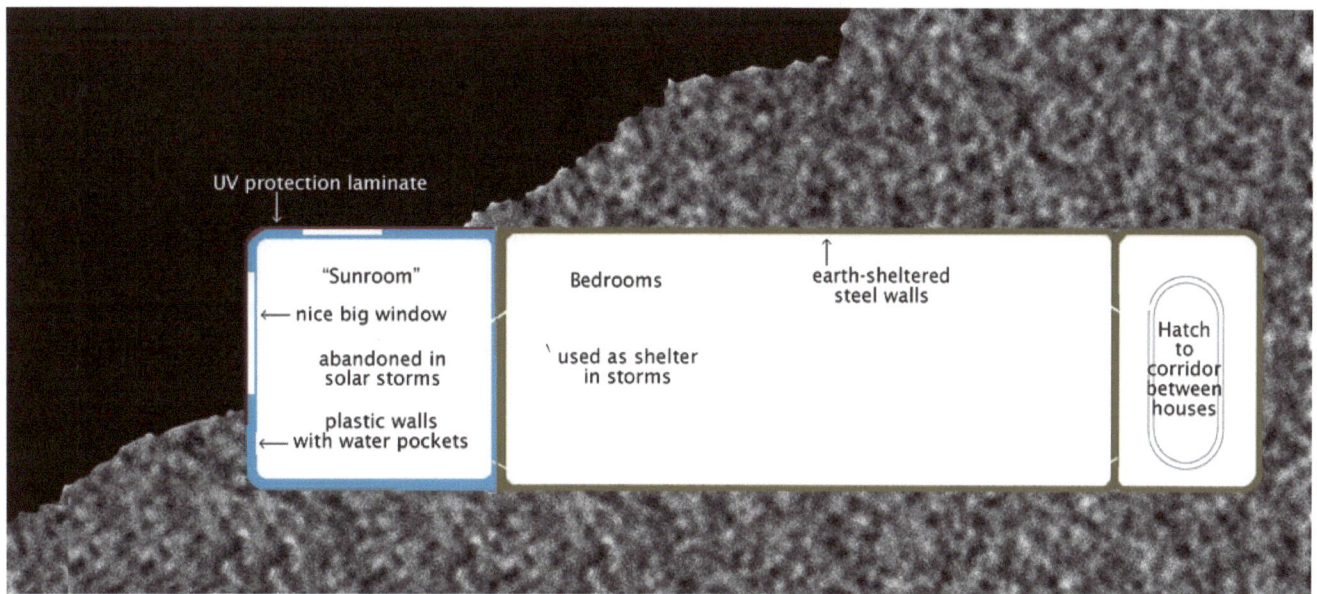

as much as they can for themselves, but computers, phones, medical equipment, scientific instruments, and many other essentials are still imported from Earth or the Moon.

Staying in touch with Earth and with the other worlds of the solar system means maintaining a space station indefinitely. The people who choose to live on the asteroid have different priorities and risk-tolerance compared to the space dwellers. Surface habitats are similar in principle to those on the station, but burying them under a thick layer of soil makes them safer: less exposed to radiation and to damage from incoming meteors. I envisage each unit having a porch – a living area with an outside view – but the main sleeping area and the corridors between units are well underground.

The whole rationale of asteroid colonies depends on our ability to tolerate very low levels of gravity. If we cannot be healthy at 1% or less of Earth's gravity, a different lifestyle would be needed, living most of the time on the station (which would spin faster for something close to Earth gravity) and only visiting the surface for mining and farming tasks.

On most asteroids, the "days" of the day/night cycle are useless for human purposes. Typically the day is between eight and twenty hours, and it is rare to have one 24 hours long. In practice, the colonists will keep the familiar days of where they came from (Earth, or sometimes Mars, in my scenario) and more or less ignore the local days. A new word would be needed for a solar day, or conversation would become very confusing. I was confused myself, planning the attack on Feronia. Again, colonists who spend most of the time in tunnels do not need to worry about this. Those involved in farming or mining would adapt.

The "years" of the orbit of the asteroid around the sun have even less significance. Typically between three and six Earth years, they have little impact on the colony, as there are no discernible seasons.

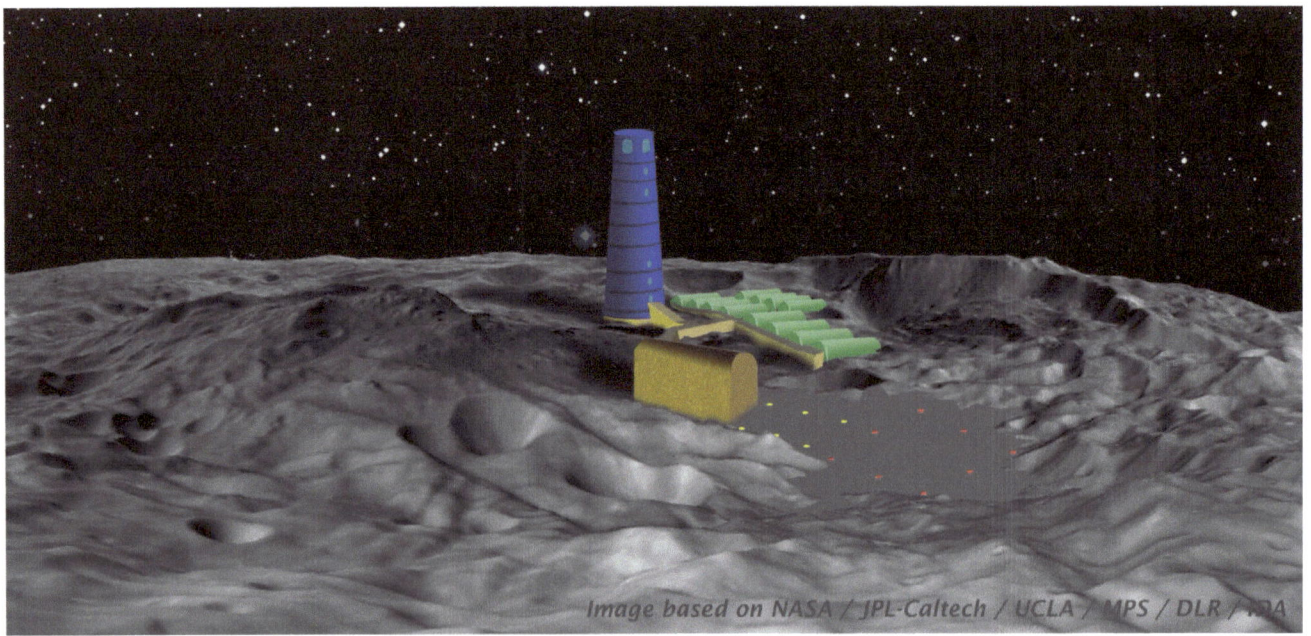

Image based on NASA / JPL-Caltech / UCLA / MPS / DLR / IDA

6.3 Martian Colonies

The Mars system presents two completely different places to live. The surface itself has gravity double that on Moon, abundant water ice, and a thin atmosphere, so living there will be easy compared to anywhere out in space. I have not included those places in my thinking. I write about space.

Living on Phobos or Deimos is like living on a very small asteroid. The gravity is extremely low; on Phobos, for example, something dropped at waist height would take more than a minute to hit the floor. But water, metals and other resources are abundant on the moons, and trade between them and with stations in other orbits is quick and easy.

Analysing the ease of travel around the asteroid belt reveals an interesting pattern. Mars orbits the Sun every two years, while main-belt asteroids take three to six years. If you want to ship something to a particular asteroid, opportunities to do so ("windows") arise more often from Mars to that asteroid than they do to any other asteroid. So the Martian system becomes the "centre" of the belt, as it can trade more easily. So, in my scenario, it is the seat of government, and a far wealthier place than any asteroid colony.

The moons are of course very close to Mars itself. Physically, about 10,000 and 22,000 kilometres. As this picture shows[9], they appear startlingly, almost ridiculously close. But in space-travel terms, they are not. It is easier to get to Phobos from Earth orbit, than from the surface of Mars. Flying from the surface takes huge amounts of fuel, 5.6 km/s of delta V: only half as hard as getting to orbit from Earth. And landing on

---

[9] Image credit: NASA / Mars Research Orbiter

Mars is actually harder than landing on Earth, as the atmosphere is too thin for convenient aerobraking. So, economically and politically, in my scenario, Mars is a separate entity, populated by independent pioneers who aim for rapid self-sufficiency, while Phobos and Deimos are much more focussed on trade, with the asteroid belt and with Earth. Of course, there is trade between to two, but people who choose to land seldom leave.

I have some design concepts for cities on Phobos, but they are not well worked out. I will have to think about it more before Lisa and her friends arrive, in *Traders of Arkady* or the numinous fourth book of that series.

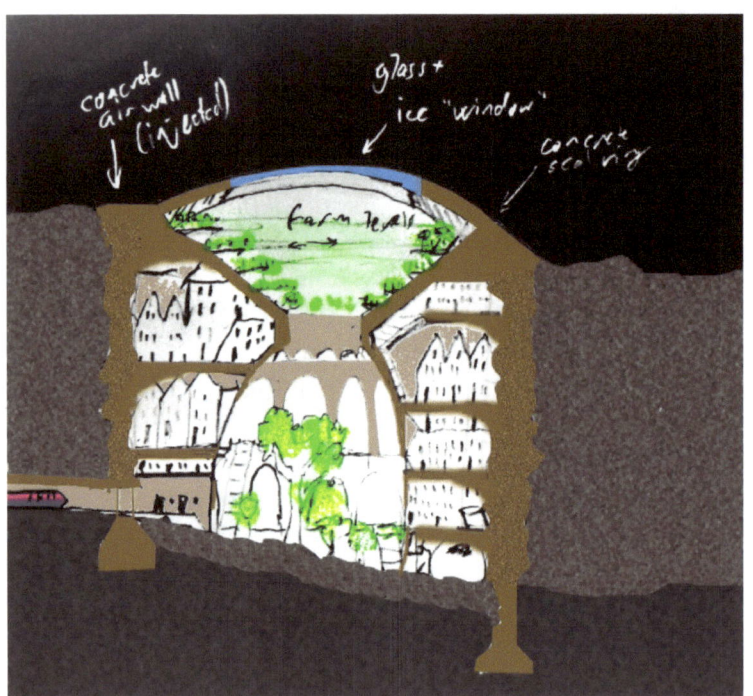

Right now, I'm thinking they will be deeply buried in the loose regolith of the surface, with dish-shaped greenhouse levels beneath an extensive dome, using ice as a structural material. With trees, of course. And cats, and people in wing-suits.

Colonies on the Martian surface naturally use a Martian calendar, where the days (referred to as sols) are about half an hour longer than on Earth. Current thinking amongst enthusiasts is to lengthen the second, but that is such a basic unit in science and engineering, so we should not mess with it. Instead, I have lengthened the minute, with just over 62 seconds in a minute. That way, we still have 60 minutes in an hour and 24 hours in a day.

There are 668.59 sols in a Martian year, so the annual calendar also needs to be different. Again, Earth-based enthusiasts have come up with a scheme I am not very fond of, with 21 months in a year, the extra months named after various people they like. I prefer to keep 12 months in the year, but have either 55 or 56 days in each month[10]. This keeps

---

[10] As adopted by Arthur C Clarke in *The Sands of Mars*. He was the most realistic early thinker about space.

the familiar relationship with the seasons, December being winter in the north, September spring in the south, and so on.

On Deimos and Phobos, they are in constant contact with the Martian surface, so they use that calendar, picking an arbitrary time-zone for alignment. Wherever the first settlement was established is the most likely one, though there is already a zero meridian assigned, *Meridiani Planum* near the crater *Schiaparelli*, and the people in my stories use that.

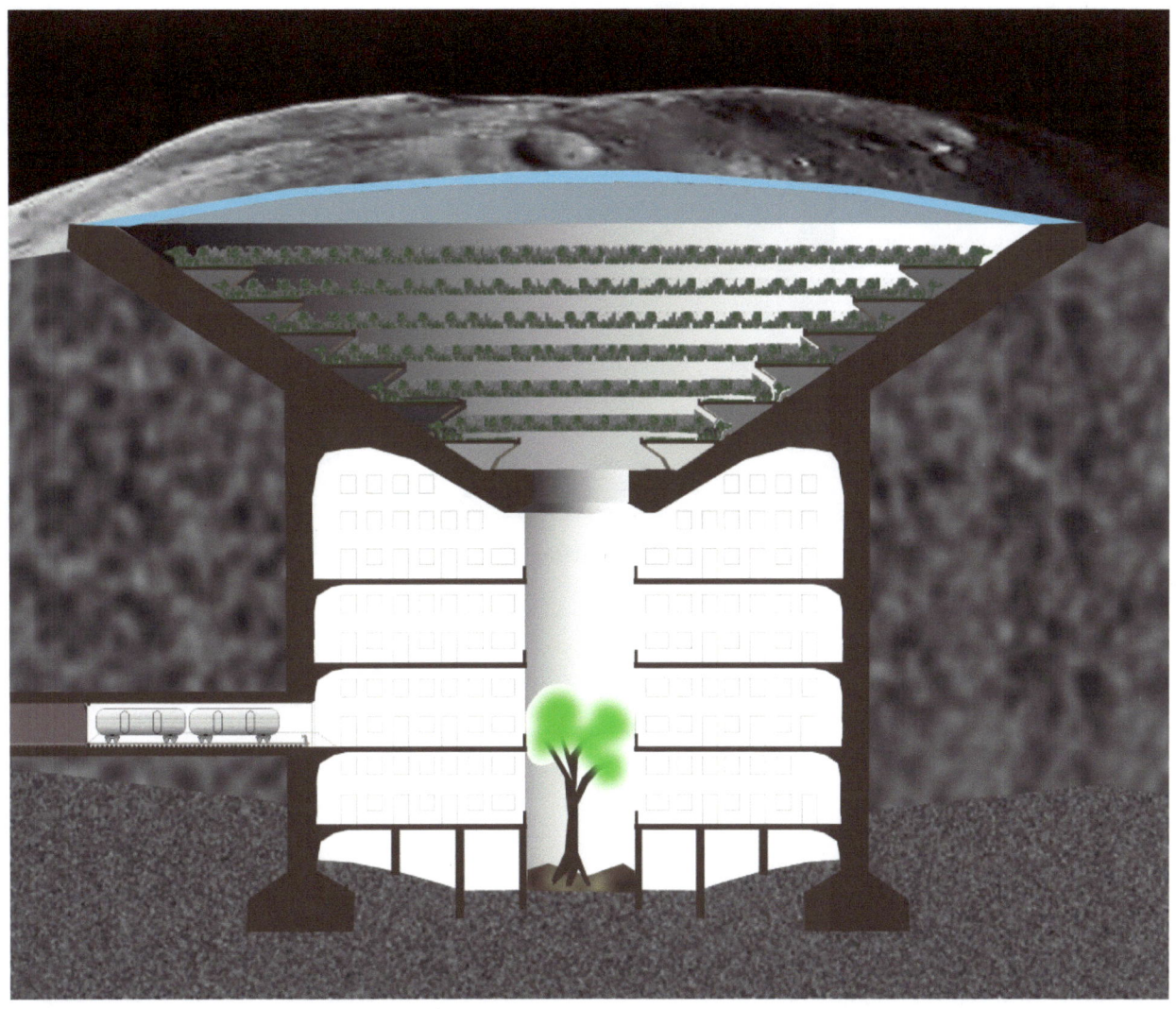

7 Exotic Equipment

7.1 Catchers

On a low-gravity body like Phobos or Deimos, find a high mountain or ridge, and put up a net. It does not have to be a very fine net, it's to catch spaceships, not fish. Attach that net to a winch, with a long, long cable. Fly your spaceship, far too fast to land on the moon, into the net. (It's on a ridge, so you can fly by if you miss.) Let the cable pay out, rapidly at first, but attach the winch to a generator or a compressor. The kinetic energy of the ship is converted to electricity or gas pressure: something you can use on the surface. Once the ship is down to a reasonable speed, detach the cable and fly calmly into port.

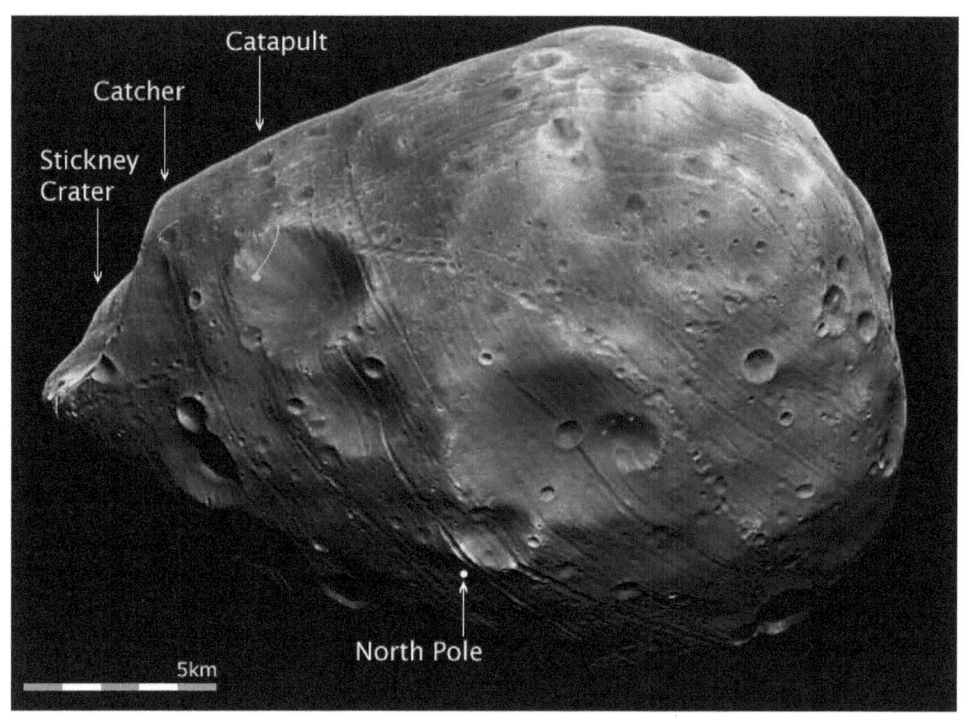

This gives your moon a huge advantage in trade, allowing cargos and perhaps even passengers to arrive from the Belt or from Earth without carrying the propellant needed to match orbits when they arrive. A hair-raising manoeuvre, with multiple gravities of acceleration and substantial risks, but a vast gain in accessibility, opening up new trade corridors. If you are very clever in siting and using the catcher, you might even be able to gradually improve the moon's orbit with it.

7.2 Catapults

In a way, the catcher gives you something for nothing, decelerating a space vessel without having to carry fuel for the delta V. A catapult does the same in reverse. It might have the same form as the catcher, using a cable to sling the load into orbit, perhaps using the same rig as the catcher. On the other hand, a simple electric railway can have a similar function. Place your ship on a railcar, using steel rails or a maglev strip, and get it up to escape velocity.

A slight hump in the tracks will send it out into space forever.

The length of the track depends on how fast you want to go, and how much acceleration you can tolerate. If we assume 5 g, that's 50m/s^2, so to escape from the Moon (2.2 km/s) you need to be on the track for just 44 seconds. Not much time, but your track has to be nearly 50km long. To escape from Phobos into a Mars orbit, you only need 12 m/s; a normal suburban train could do it, maybe even a bicycle. In reality, the useful thing to do is to escape the Mars system, so you want something at least as powerful as the Moon catapult.

One of the nice things about catapults and catchers is that they can be helpful even without doing the whole job. A moon catapult can save a ship some of the fuel needed to get to orbit, before it is built out far enough to get escape velocity, or a Phobos catcher can save you a fraction of your deceleration burn.

Such things have been proposed for Earth, but unless they are on an Everest-sized mountain they run into the problem of air resistance. If you enclose them in an evacuated tunnel, like the Hyperloop, you have the problem of what happens at the end of the tunnel, when you release the capsule into air.

7.3 Tethers

There are numerous ideas for attaching very long tethers to things, thousands of kilometres in many schemes. If you are in a low orbit around a planet or moon, a weight on a tether can give you fractional gravity from tidal forces, so you do not need to spin your station. If you attach a tether to a station and then spin up the whole thing, you can use it as a space catapult, accelerating outgoing cargos using station power, or having your tether dip down to the surface and pluck things into orbit. I have not included such things in my thinking, as they seem complex and dangerous compared to rockets.

7.4 Space Elevators

The ultimate tether system is one attached to a moon or planet, tens of thousands of kilometres long, providing a free ride to space. I would put these far in the future, if ever. An Earth-based elevator requires materials that are just unfeasibly strong for their weight.

8 Design Considerations

8.1 Gravity

From experiments aboard the ISS, we know that long-term exposure to free fall (or microgravity as NASA misleadingly calls it) causes health effects. In the near term, digestion and bowel movements are disturbed by difficulty in separating gas from liquid. Those prone to motion sickness feel that, and there is a sensation of panic from constantly falling. These effects settle down after a few days or weeks.

In the longer term, there is bone loss from the lack of constant load, which also leads to high levels of calcium in the urine. Exercises, such as running on a treadmill while held down by bungee cords, seems to help with this. In space, you do not need strong bones, so it may be that space dwellers will simply live with the effects, and only worry about them if they need to go into a gravity well. There are also effects on the eye due to changes in the distribution of fluid around it, but these seem to be correctable with glasses.

What we do not know, because no experiments have ever been done on it, is the long term effects of moderate gravity, between the 0.05% you would get on Phobos to the 38% on Mars. People I have spoken to are fairly confident that Lunar or Martian gravity are healthy, but the area in between is completely unknown. It would be relatively simple to put up a spinning station, or even one with a long tether to provide tidal drag, but because space administrations are not serious about travel to deep space, they have never looked into it. Animal experiments would be a good start.

My characters have additional worries about gravity; in addition to living in reduced gravity for long periods, they also plan to have babies there. We have absolutely no knowledge of how that would go. In my books, small children and cats are introduced to low-gravity habitats (10%) in the first years. The cats are allowed to breed first, and the outcome of that experiment is eagerly awaited. From the earlier books, set later in time, it is clear that it went well, and women start becoming pregnant soon after. Because of the radiation environment, conception is always by IVF, using stored sperm and eggs.

If my assumption turns out to be wrong, and Martian or higher levels of gravity are needed for health, a number of things would need to change. Stations and long-distance ships would need to spin faster, and they would be correspondingly heavier and less able to get around. Surface colonies would be occupied only for short periods around specific tasks, with permanent residence aboard spinning stations. It would not make settling in the Belt impossible, but it would not fit with my stories.

Another possibility is that drugs, or ultimately genetic engineering, would mitigate these effects. Then a special breed of humanity, maybe even looking physically different, would occupy space. A great story, but not the one I have been writing.

8.2 Radiation

Earth's atmosphere protects us from most of the dangerous radiation that is flying around in space. The atmosphere keeps out much of the ultra-violet (UV), x-ray and gamma rays. On Earth, we are exposed to more of these if we climb a mountain, or particularly if we fly in a plane, but the risks are known and we find them acceptable.

In space, UV in sunlight is extremely strong, including high-energy UVC rays that are not found on Earth. Our travellers can protect themselves from these by simply staying inside, and by using small boats instead of spacesuits whenever possible.

There are two additional hazards, however, that spacers cannot completely protect themselves from. **Solar storms**, the disturbances on the face of the Sun that cause the Aurora Borealis and Aurora Australis, are streams of ions and protons: charged particles that can cause genetic damage and cancer. Thick layers of shielding can protect from these, but it needs to be on the order of a metre to be effective. Various emergency measures are used for this in the books, and all the ships have at least 10cm of shielding. **Cosmic radiation** is more of a problem. This comes from all around the galaxy, so it is not directional, and it is extremely high in energy. Many of the particles are more energetic than anything we can produce on Earth. Astronauts report visual effects, tracks of particles in their eyeballs like the cloud chambers at CERN. Nothing we can do in the way of shielding can stop these. There is some suggestion that metal shielding, of the kind used in present-day space habitats, does more harm than good. High-energy particles hit the metal and cause cascades of lower-energy gamma and x-ray radiation, leading to more problems than the original particle might have done. With this in mind, I use plastic for the walls of my habitats. NASA is now experimenting with expandable modules with "composite" walls, perhaps for this reason.

8.3 Pressure

Humans need air to live. If you are suddenly in vacuum (what my characters call the "Sucking Dark"), you will pass out in thirty seconds or so, your brain will be starved of oxygen, and you will be irretrievably dead within four minutes. Earlier stories about your blood boiling through your skin have been disproved, so there's that.

We need the air to contain oxygen, at a partial pressure about the same as on Earth. We can tolerate about half that, though various degrees of altitude sickness arise. Some early space habitats used pure oxygen at a lower pressure (about 250 millibars), which is OK for human breathing but an extreme fire hazard. The three men who died in the Apollo 1 capsule were victims of that decision. Modern space habitats, and the ones in my stories, use 75% nitrogen and 25% oxygen, with a little less than a bar of pressure. It seems to suit us.

8.4 Rocket Basics

Three factors make space travel difficult: propellant, energy, and time. To make a rocket go, you need to throw something out the back very fast. That is what we call **propellant**; it makes your ship accelerate forward. If you throw it faster, you get more acceleration. The throwing speed is called the **exhaust velocity**, measured in metres per second. The amount you throw out every second is the **throw rate**, in kilograms per second. The rocket pushes your ship forward with a force known as **thrust**, which is measured in kilo-Newtons.

Thrust = throw rate x exhaust velocity

Divide by 100 if you want kilograms-force. Acceleration is the rate of change of velocity (remember calculus), and

Acceleration = thrust ÷ mass of ship

If you want to take off from a planet or a moon, your acceleration must be greater than gravity, otherwise you will just sit on the ground.

If you keep your rocket going long enough, you will run out of propellant. If you have planned your journey correctly, this will not happen before you reach your destination; otherwise, you drift in space forever. Generally, spaceships need large fuel tanks, something else the movie makers miss. The overall performance of a ship, its ability to make a specific trip, is measured by **delta V**, simply the total of all the acceleration it can manage before running out of propellant. This is a velocity, the change in speed across the whole trip, in kilometres per second. This is governed by the "rocket equation," derived by Tsiolkovsky in 1903.

$$\Delta V = V_e \times ln\left(\frac{m_0}{m_f}\right)$$

This says that delta V is the exhaust velocity multiplied by the logarithm of the **mass ratio**: the initial mass of the fully fuelled ship, divided by the final mass, when the tanks are empty.

get from carrying more and more fuel. Even if 99% of your take-off mass is propellant (like a Saturn V rocket), your final velocity is only 4.6 times your exhaust velocity. Propellant is lighter than structure, so a vessel like that is virtually all tanks.

8.5 Propulsion

Anything that moves in space needs a rocket to make it go. There three basic choices, chemical, thermal and ion.

A **chemical** rocket (the only kind you ever see take off from Earth), burns a **fuel**, such as hydrogen or petrol, just like a jet engine. With no atmosphere, it can't burn the fuel in air, so it needs a supply of oxygen. This is called an **oxidant**, usually simply liquid oxygen. The small boats in my stories use chemical engines, because they are simple and relatively easy to build.

A **thermal** rocket applies heat to a liquid or gas, and throws the stuff out the back simply using vapour pressure. This is not very efficient, but it has the advantage that you can use any kind of material as propellant. My boats use steam thrusters, because they are safer to work around than chemical rockets.

An **ion** rocket is a particle accelerator; it ionises a gas to turn it into electrically charged plasma, then uses magnets to accelerate it to high speeds and throw it out the back. Sending out all those positive ions would rapidly charge up the ship to a high electrical potential, so

Mass ratio	1.111	1.250	1.429	1.667	2.000	2.500	3.333	5.000	10.000	20.000	100.000
Delta V	0.105	0.223	0.357	0.511	0.693	0.916	1.204	1.609	2.303	2.996	4.605
Fuel as % of mass	10	20	30	40	50	60	70	80	90	95	99

This table shows the absolute tyranny of the mass ratio, and the diminishing returns you

there is an electron beam out the back, to meet the ion stream and neutralise it. Ion rockets have very impressive exhaust velocities, but it is hard to get a decent throw rate, so acceleration is very small. I am assuming a much higher throw rate than anything feasible right now, so this is one area where my extrapolation of present technology is rather extreme.

8.6 Power

Throwing things away quickly takes energy. With a chemical rocket, the burning of the fuel in oxygen is all you need. For thermal or ion rockets, though, you need an external source of energy. Using chemical fuel makes no sense; if you were going to do that, you might as well have a chemical rocket. So the choices are solar power, or nuclear fission. Nuclear fusion will be great if they ever get it built, but I have lost faith in that. Solar provides limited amount of energy (even less if you go out to Mars or the Belt) and uses big high-tech silicon cells, which might be hard to make off Earth. Nuclear fission just needs big chunky steel and similar hardware, produces plenty of power, and the technology is well established. The environment is much less of a concern in space than on Earth – there is no atmosphere to pollute, and there is already a fair amount of radiation, which a reactor accident would just be adding to. A malfunction on a ship would probably kill you instantly, but losing your main source of power on a spaceship will kill you anyway, so it's just a matter of time. In my stories, I use nuclear fission as the primary source of power on ships and stations. That is used to create chemical fuel from local materials (mostly ice) to power the hold, boats and other mobile units. Aboard ship, the reactor generates steam for conversion to electricity, powering the main drive and anything else aboard. The reactor would run hot, using molten salts or something similar for heat exchange, so the heat from that can be used for smelting metals and perhaps even glass, directly in the reactor building.

No energy source is 100% efficient, so there will be waste heat. ES *Shoemaker* has some impressive vanes to dissipate this; probably the stations in the earlier books should have had them too.

In *The Dark Colony*, I needed a power source that could travel, but was lighter and smaller than a reactor, for the original *Dancer*. For that, I used a Radioisotope Thermal Generator (RTG) similar to what they used on the Moon in the Apollo program (though larger). This is just a

sub-critical lump of Plutonium 238, which gets hot enough to run your power source simply by being radioactive. At present, Pu238 is hard to come by because you can make atom bombs from it. However, it is made in fairly simple nuclear reactors that can also generate power, so it should be easy to come by in space.

8.7 Local Navigation

The characters in my books spend a fair amount of time flying small boats, and some explanation of how this is done may be in order.

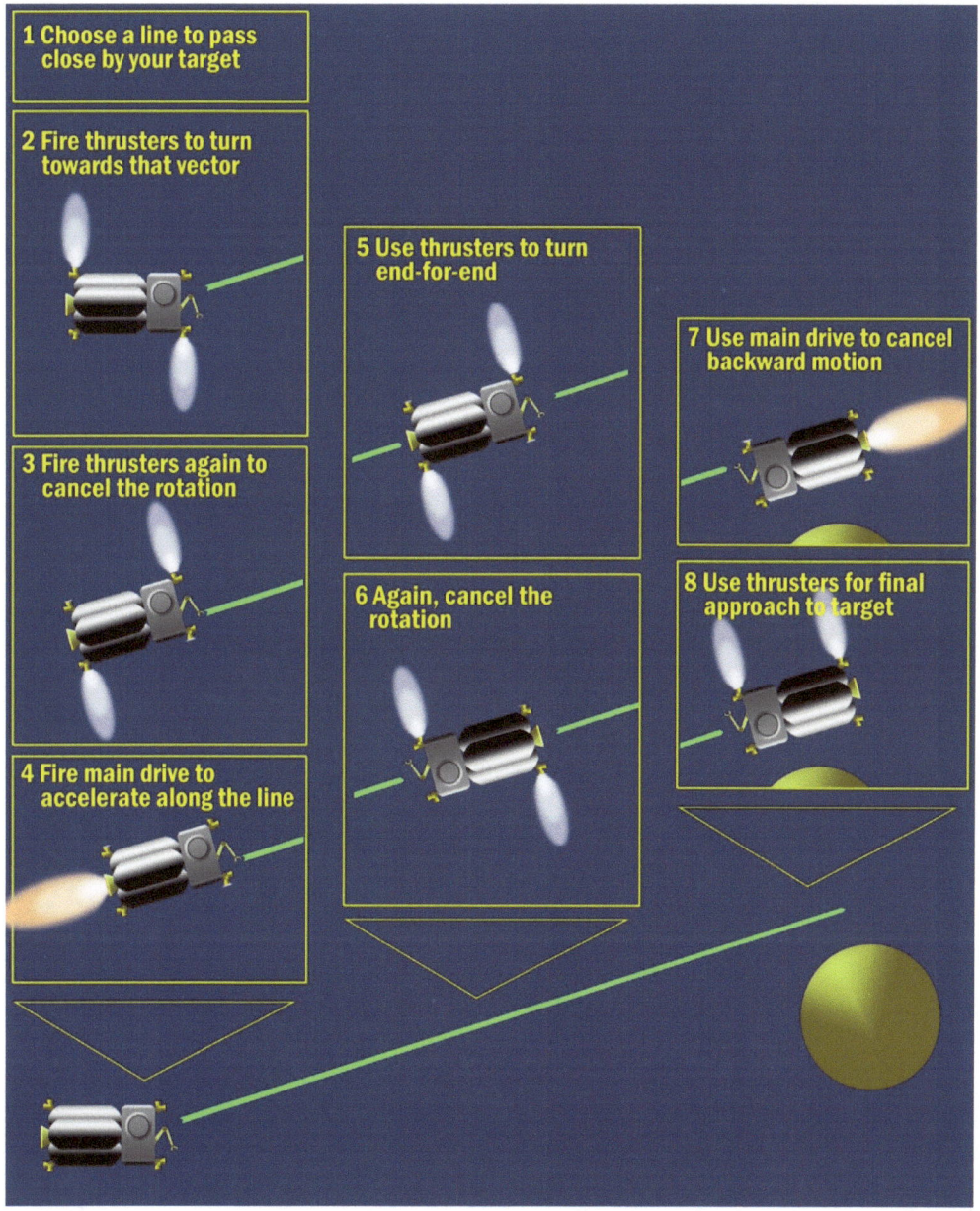

For motions of a few metres, the thrusters are enough. For longer movements, you point the boat in the direction of your target and use the main drive. Half way to the target, you turn through 180° and fire the drive again, aiming to be stationary when you get to the target. Bearing in mind that a rocket is a great big blowtorch and all the habitats are made of plastic, you aim a little to one side, cancel your forward motion with the main drive, and use the thrusters for the last few metres.

Movements of short duration and distance are done by eye, without extensive calculations. Longer journeys, say to a different orbit, or to match with a passing station, are calculated by the computer. However, the pilot stays in the loop, doing the physical movement of the boat, just like on an aeroplane. This approach, derided as "meat servo" by the proponents of automated machines, is the one my characters choose because they feel safer with it, whatever Earthlings prefer. Never trust a robot with your life.

A co-ordinate system is needed for pilots to talk to each other and to the computers, describing directions and positions in space. The basis of the system is the largest object in the surrounding space. Around a boat, it is forward/ aft, port/ starboard and zenith/ nadir.

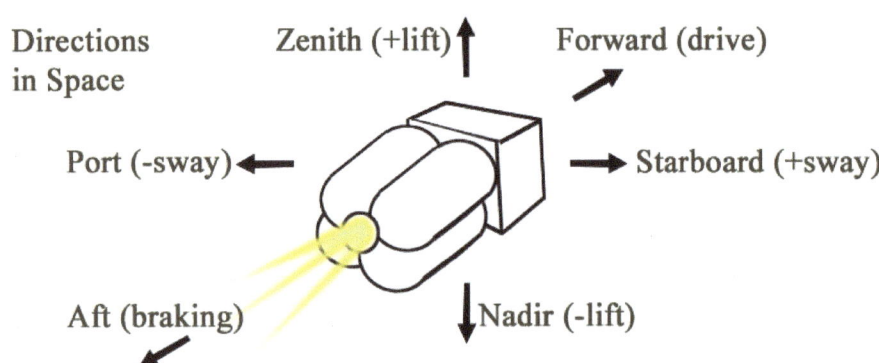

Directions in Space

In the space around a station, it is east/ west (around the orbit), in/ out (towards or away from the planet) and north/ south (in line with the poles of the planet). Generally, all boats and ships keep their zenith directions aligned, making turns horizontally (in yaw) rather than tilting up and down, wherever possible. As a result, zenith is often referred to informally as "up," even though there is no gravity.

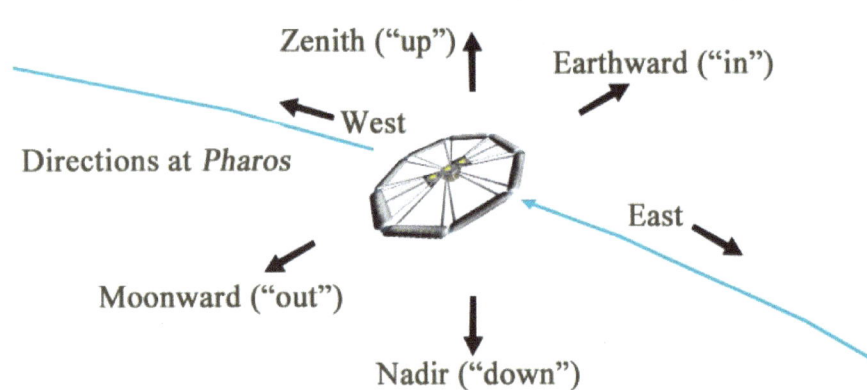

Directions at *Pharos*

Coordinate systems on ships are complicated by the tumbling or spin of the ship, so a static and a rotating frame are needed for different purposes. The characters in *Freedom at Feronia* discuss this at some length, coming up with different terms for the spinning frame and the static frame.

9 Solar System Travel

9.1 Earthspace

I coined the word "Earthspace" to refer to the Earth-Moon System: The Earth, the Moon and all the artificial satellites within its gravitational sphere. In my books, it's an independent nation, with the same status as any on Earth.

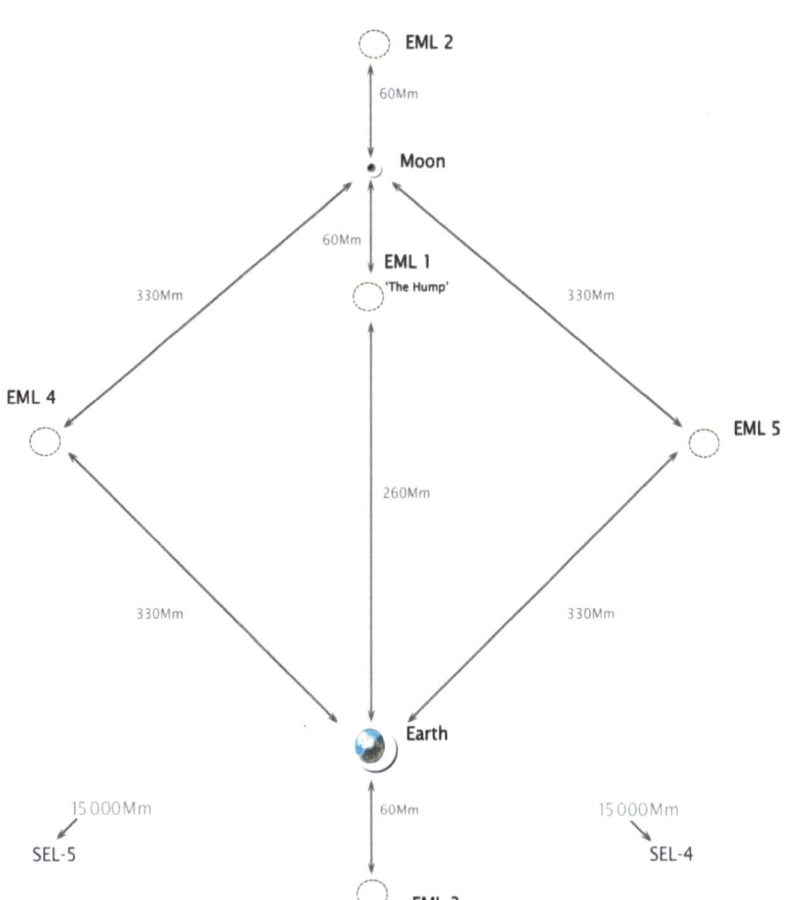

Science fiction has tended to concentrate on EML-4 and EML-5, because a habitat placed at one of those points is stable forever, with no need for station-keeping thrusters to keep it in place: good for permanent colonies aiming for self-sufficiency.

However, my characters like to trade and to import materials from the Moon, and for that EML-1 has much more potential. You have to spend some energy station keeping so you don't fall into the Moon or back to the Earth, but you are only 16 hours from the Moon, instead of the 4 days it takes to get to EML-4 or EML-5. So those points will be inhabited by crazy O'Neill colonists, not proper spacers.

EML-2 is the best place for launching missions to Mars, so there should be a station there before long. EML-3 is perhaps good for science, but not a sensible place to live.

The Sun-Earth Lagrange points are even more remote; people do not live there in my stories.

This chart shows the delta-V costs of travelling to the various points in Earthspace. Note that the cost is the same whichever direction you are going, unless you are landing on a planet with

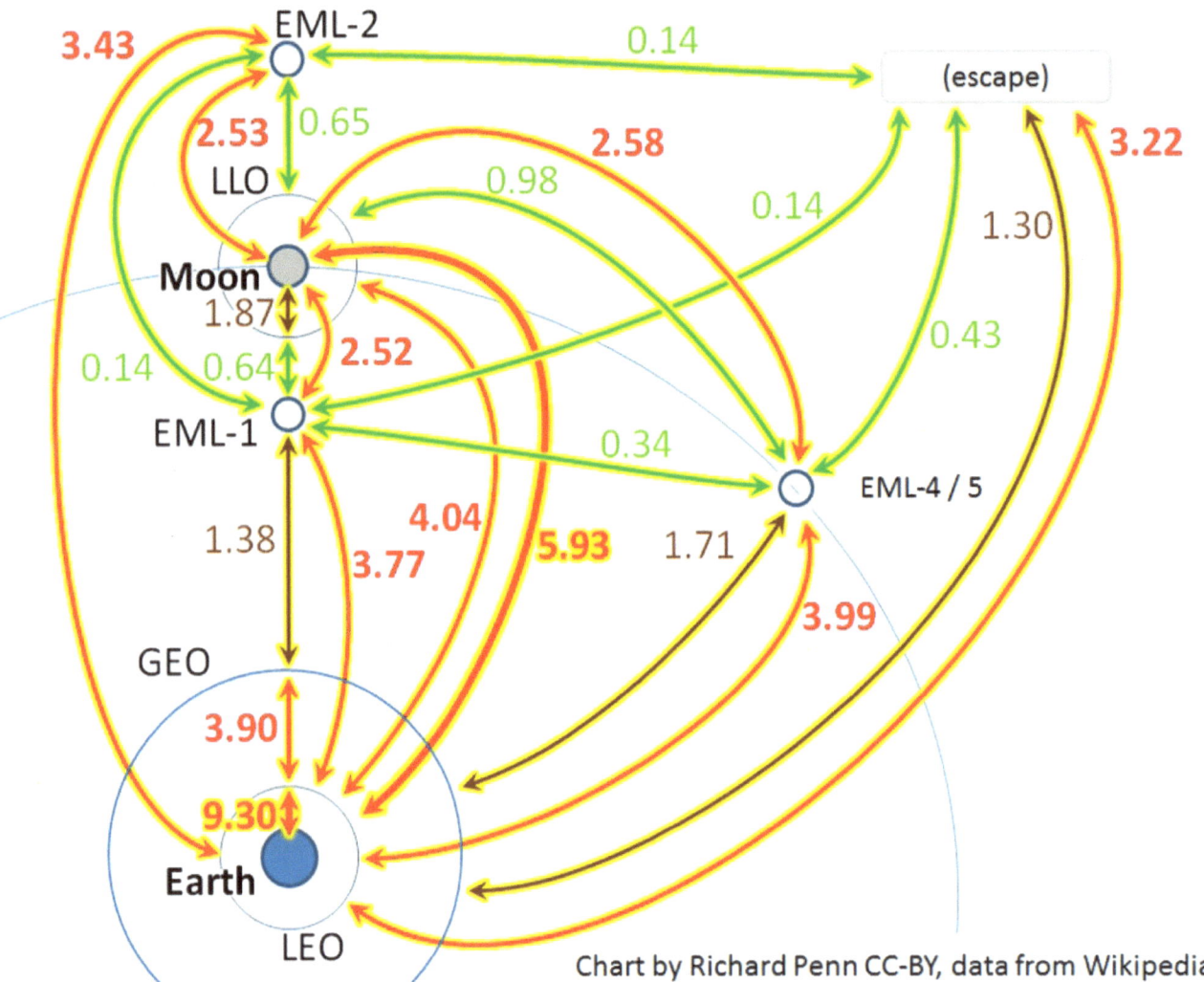

Chart by Richard Penn CC-BY, data from Wikipedia

an atmosphere, but we don't do that. Notice that a ship capable of just 1 km/s of delta-V can do some useful trips among the EML's and down to Low Lunar Orbit (LLO).

Some of the journeys shown here take a very long time, and a ship may choose to expend more fuel to get there quicker. This is particularly true of L4 and L5; the delta-V's shown here are for transfers that take up to two weeks.

Other useful orbits in Earthspace are: Low Earth Orbit (LEO) where the International Space Station and many simpler satellites are; Geosynchronous Orbit (GEO) where a satellite can maintain a fixed position over Earth's equator (good for comsats); and the Sun Synchronous Orbit (SSO), a near-polar orbit that passes low over the same territory every day, good for spies and the military.

There is an official parking area called the graveyard orbit just outside GEO, where comsats are supposed to go when they are no longer operational, but many broken or expired ones are in other near-synchronous orbits. In my stories, this was a happy hunting ground for salvage operations, in the early days of the occupation of space.

This chart shows the number of satellites and pieces of wreckage in various orbits right now[11], in 2016.

Geosynchronous Transfer Orbit (GTO) also has a lot of defunct hardware in it. This is a highly eccentric orbit starting at LEO and reaching up to GEO. A non-functional satellite may be left there, as there is no point in continuing with its mission, or booster problems may prevent it entering GEO.

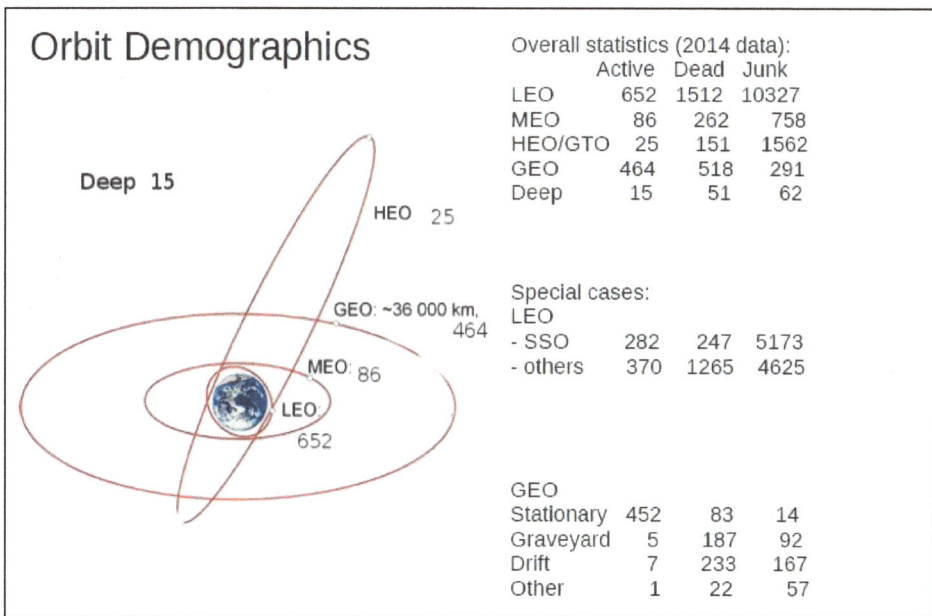

Orbit Demographics

Overall statistics (2014 data):	Active	Dead	Junk
LEO	652	1512	10327
MEO	86	262	758
HEO/GTO	25	151	1562
GEO	464	518	291
Deep	15	51	62

Special cases:			
LEO			
- SSO	282	247	5173
- others	370	1265	4625

GEO			
Stationary	452	83	14
Graveyard	5	187	92
Drift	7	233	167
Other	1	22	57

---

[11] Illustration courtesy Jonathon McDowell, who has much more information at http://planet4589.org/talks/global/global16.pdf

9.2 Travel to Mars

As the "escape" arrows show in the chart on page 70, once you are high in the Earth's gravity well it costs very little to leave it altogether, and go into orbit around the Sun. Pick the right time and direction, and you can get to Mars or a nearby asteroid. Shipbuilding and fuel stores need to be at points like EML-1/2, so you can start your interplanetary journey with as much fuel and supplies as possible.

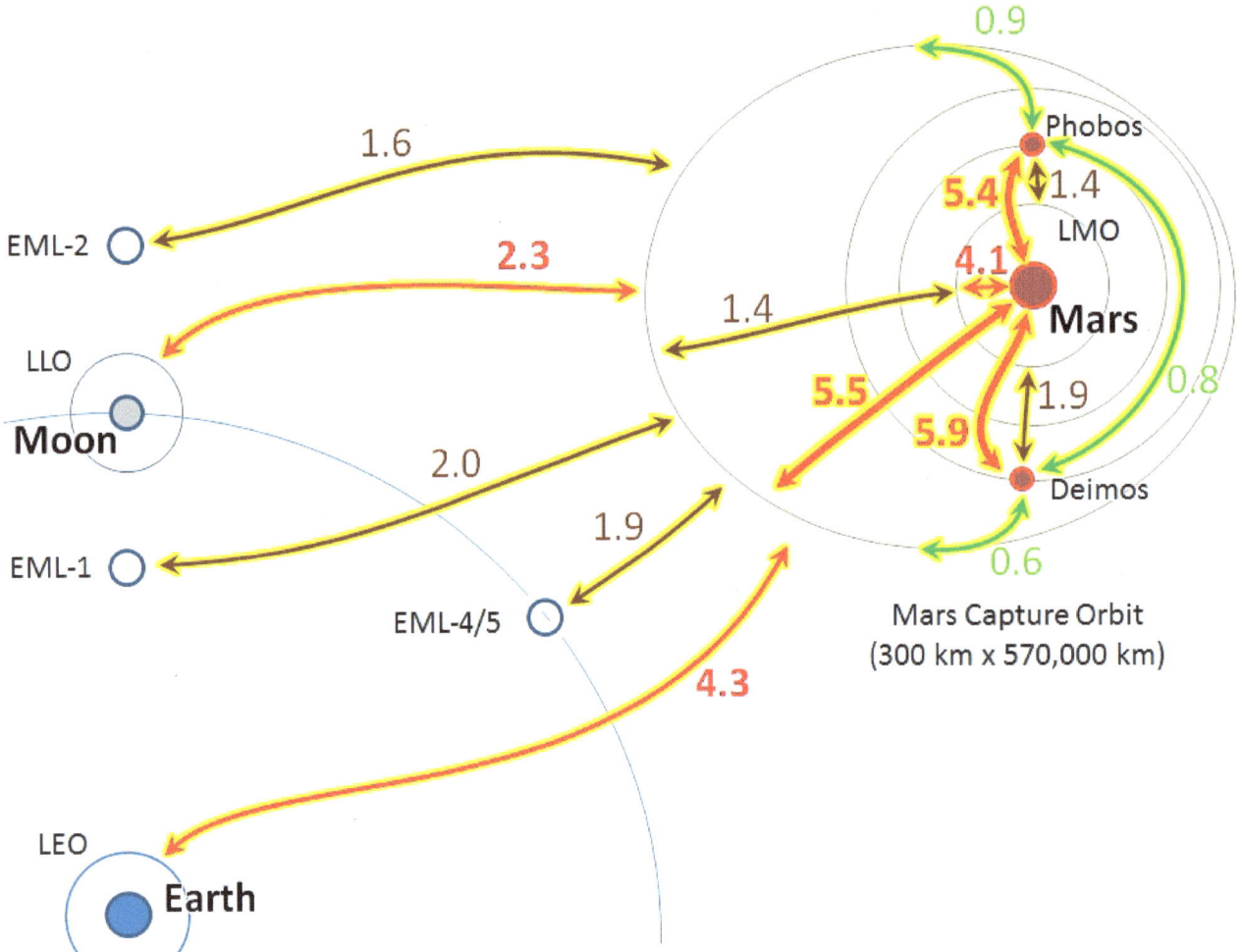

Mars Capture Orbit
(300 km x 570,000 km)

This chart is for transfers to and within the Mars system[12]

---

[12] Chart by the author, based on data from Hollister "Hop" David

It is clear from this that a fairly modest ship can get you to the Mars system, and especially to Phobos and Deimos. Landing is hard, but that is always going to be the job of a specialist landing vehicle, something with massive heat shields, and either a parachute or some very specific landing thrusters. If you plan to take off from Mars, the launch vehicle is also very specialised, mostly fuel tanks.

This is why a there-and-back mission to Mars is such a bad idea. You need to start with a ship capable of getting you into the Mars system; not a huge challenge, as we have seen. Anything that can land on the Moon can do it. But along with that ship, you have to carry a lander, with that massive heat shield and all the propellant needed for landing. And a launch vehicle, with all the fuel it needs to take off. Then you can rendezvous with your interplanetary craft, which also needs to contain enough fuel to get back to Earth.

In my scenario, you start from a base at EML-2, carrying machines and people and enough fuel to get you to Phobos, similar to a Moon landing. At Phobos, you get out of your ship and get into a lander, built there from local materials. It carries you down, jettisoning the heat-shield on the way, and lands you among the Martians. If you should foolishly wish to go back, you buy a launch vehicle from the Martians (humans, not little green men), which they have built from local materials. Your interplanetary ship has been refuelled from the ice mines of Phobos, and you can happily fly it back. It is called In Space Resource Utilisation, and it is the only way to fly.

Note that all the orbiting places in the Mars system are linked by arrows with 2.0 km/s or less. This means that any of the small boats pictured earlier could fly you around the system, and the tugs could take a ten-tonne load anywhere in a day or two. This means that stations in Low- and High-Mars orbits, and on the moons, can be in constant contact and trade. This, combined with relatively easy links to the asteroid belt and to Earthspace, means that these stations are at the heart of the solar system, as far as trade and travel are concerned.

http://hopsblog-hop.blogspot.co.uk/2013/04/cartoon-delta-v-map.html

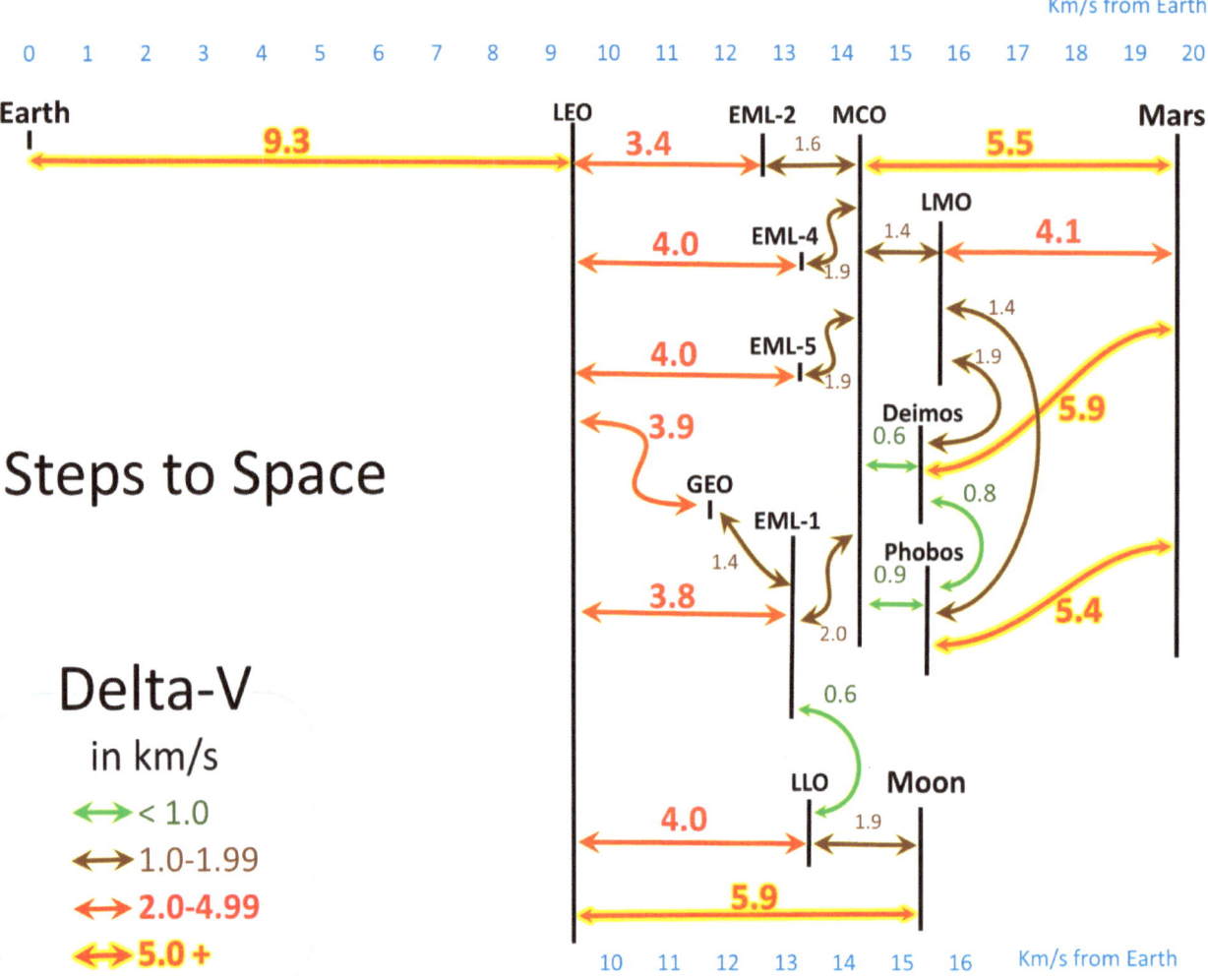

Chart by Richard Penn CC-BY, data from Hollister "Hop" David: hopsblog-hop.blogspot.co.uk

It is worth emphasising that all the numbers in these charts are minimum values, and a lot of things have to be lined up to achieve them. Windows for transfer between Earth and Mars only open up once in 2.1 years; try to travel any other time and the Delta V is orders of magnitude higher. Also, these figures depend on using high-thrust motors, so that the changes in velocity are only made low in an orbit. Because of the "Oberth

9.3 Mars Cyclers

Once travel to Mars becomes established, it will make sense for stations to be placed in Mars Cycler orbits, fixed orbits which regularly bridge the two orbits, providing a protected transfer between the planets. One example of these is shown here.

The story begins with a conventional economy transfer, coming close to Earth on day 110, and dropping off at Mars 30 weeks later, on day 320. In subsequent years, the station continues to coast between the two orbits, but at those times the planets are differently aligned.

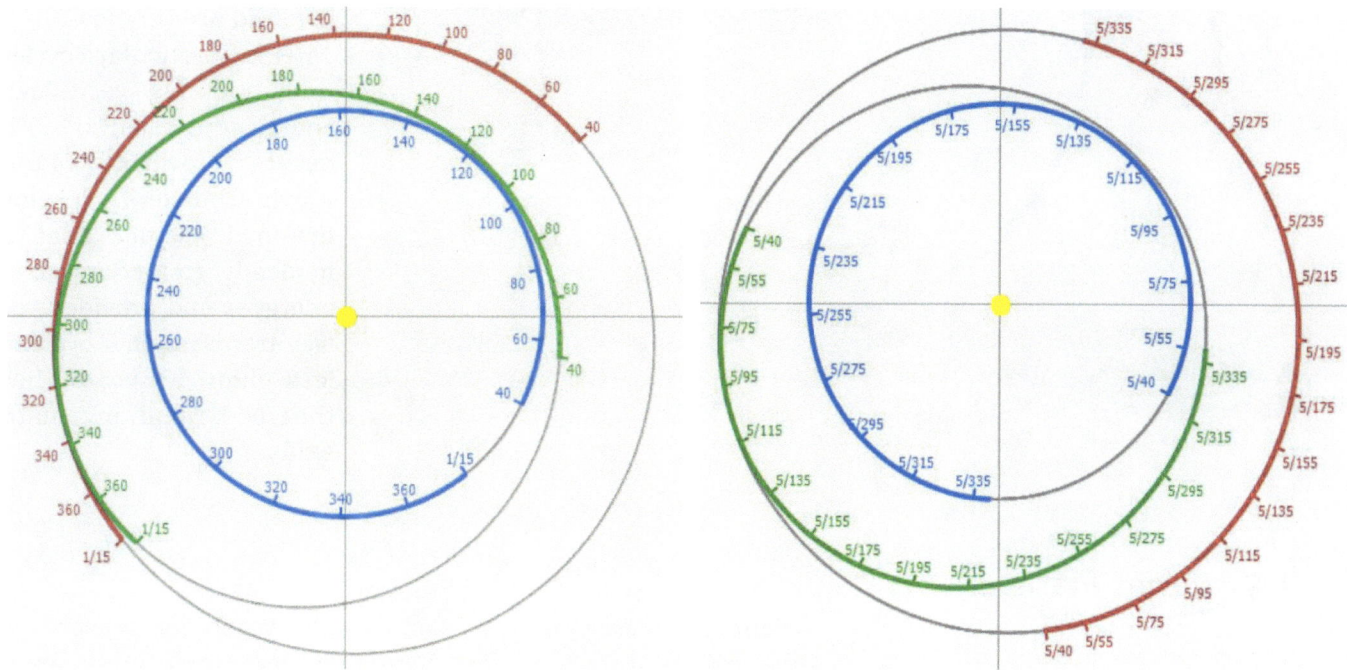

For example, in year 5, shown on the right, Mars is on the other side of the Sun when the cycler meets its orbit.

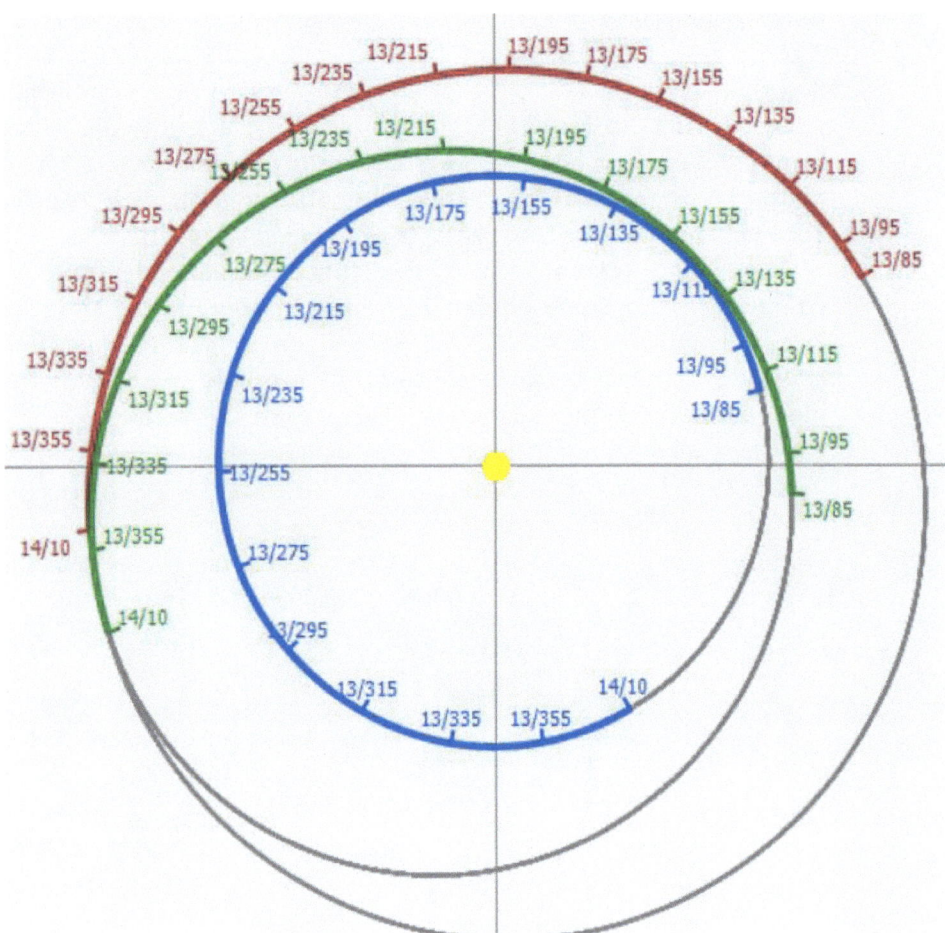

However, by year 13, almost exactly 7 Mars years have passed, and the cycler has passed through close to 9 orbits, so the situation repeats itself, allowing another Earth-Mars transfer. The match is not exact, so small adjustments to the cycler's orbit would be needed, but the thrusts involved are very low.

This particular cycler orbit requires very low-energy adjustments, but it never provides Mars-Earth transfers. In my fiction I assume that a modestly greater input of energy would provide two-way transfers on a cycle of less than 13 years, but that is beyond my math skills.

9.4 Other Planets

Venus and Mercury are both within reach for reasonable ships. I have not worked out the details, but Mercury is a great source of metals and other minerals, it is just extremely hot and exposed to solar radiation. A crater near the pole in perpetual shadow might be a viable site. Venus is ridiculously hot, and with the high atmospheric pressure it is a huge challenge to get to the surface. But, if you fill a balloon with ordinary air, it can float in the atmosphere at a level where the temperature is quite reasonable. The atmosphere has all the gases you need, and perhaps other materials in the dust.

The outer planets are just exceedingly hard. It takes many years to get to the Jupiter system with the kind of technology shown here, and then you enter the system at nearly 60 km/s! The probes we send there do dozens of gravity-assists, taking many more years to kill that velocity, all the while exposed to gamma- and x-radiation many times that in the inner system. I would never say it is impossible to go there, we are going to the stars after all, but it will not be with anything we presently know how to build.

9.5 Transit Asteroids

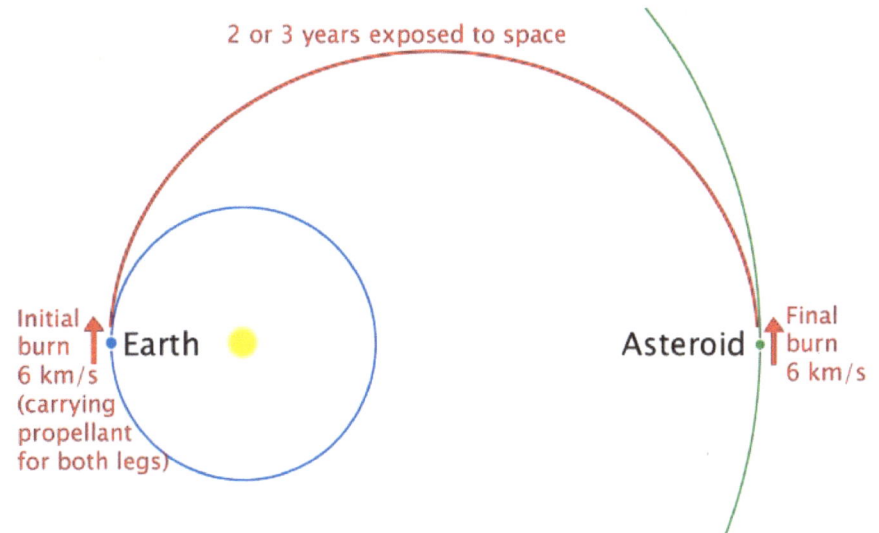

The concept of the transit asteroid is central to my books and to the scenario behind them. There are two problems with long journeys in space: fuel and radiation. The problem with fuel is that your rocket throws a lot of it out the back, in order to make your ship go forward. You need to accelerate away from your starting point (Earth or Mars), and then when you get where you are going, you have to accelerate again. Each time, you use a lot of fuel, and the period in between is about half the period of your target asteroid, so one or two years, at least. While you are coasting, you are soaking up radiation, from the Sun during solar storms, and all the time from deep space, in the form of high-energy cosmic radiation.

The transit asteroid addresses both these problems. If you can find a small asteroid (there are millions out there) that happens to pass close to where you are, and then later in its orbit passes close to where you are going, you can hop aboard. If the transit asteroid passes within say 4 million kilometres, and has a relative velocity of less than 4 kilometres per second, you can make the hop in a few weeks, with about 7 km/s of delta V. Once you are on the little asteroid (500m or smaller would work), you can bury your habitat in the loose surface material for radiation protection, and you can mine the rock for propellant. This only works with a nuclear rocket, as that can use inert materials for propellant. We are not going to find chemical rocket fuel lying around anywhere.

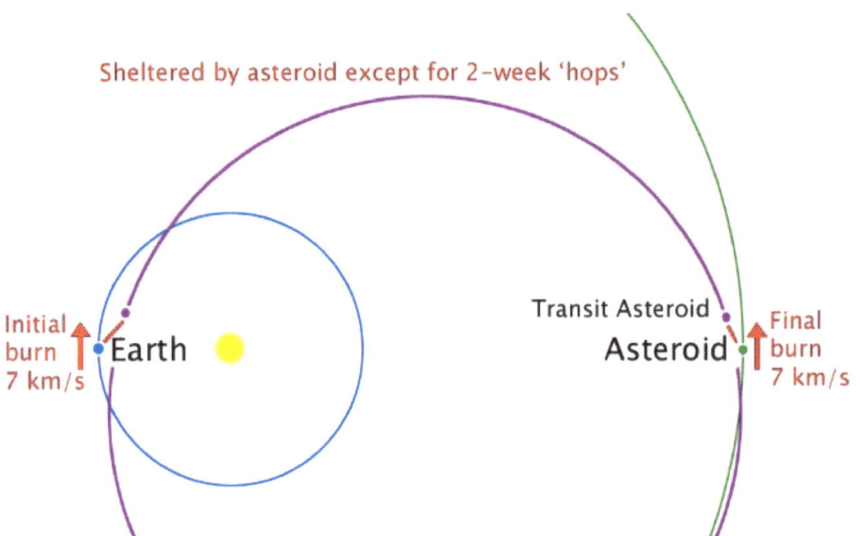

So, you spend your year or two living on the transit asteroid, waiting for it to pass close to a big asteroid in the main Belt. When it does, you hop off, again taking only a few weeks.

Asteroids like 8013 Gordonmoore, which pass close to the Earth, are relatively rare, providing transit opportunities only once in a few years. But from Mars, several times a year, asteroids pass close by that are headed out to various places in the Belt. And that is only counting the ones we know about; there are probably many more small rocks passing between the Belt and the orbit of Mars. You could almost pick a specific destination and wait for the next one to roll by. In reality, any asteroid that passes close to Mars will pass several destinations in the Belt. In the years I am writing about, there are not enough permanent colonies of hundreds of people setting out to exploit every one. But prospectors and mining companies will be looking to get to every accessible rock, in hopes of finding that mineral bonanza they all dream of.

Vessels and Stations of Earthspace and the Belt — Richard Penn

My whole model of Belt colonisation is based on the movement patterns shown here. Each slanted line represents a known asteroid that could be used to transfer between these

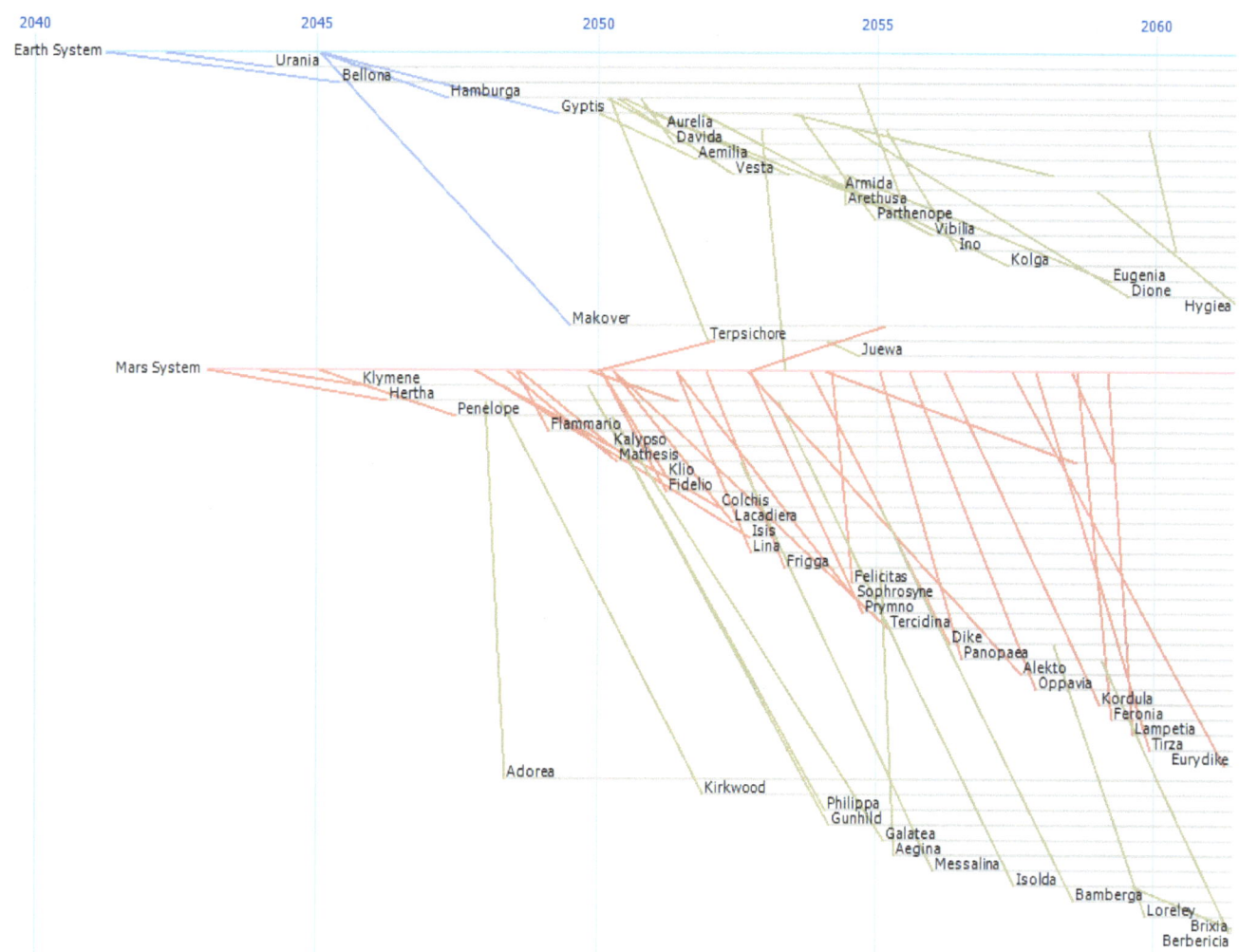

places. Note that Terpsichore is the only rock served from both Earth and Mars; that is why I set the story there.

The model is based on a simplified version of orbital mechanics, so in its exact details it is probably wrong. But I believe the general pattern is realistic, that something like this could really happen.

9.6 Colonisation Scenario

This is the long-term structure of human settlements in the system, by the time of *The Dark Colony*.

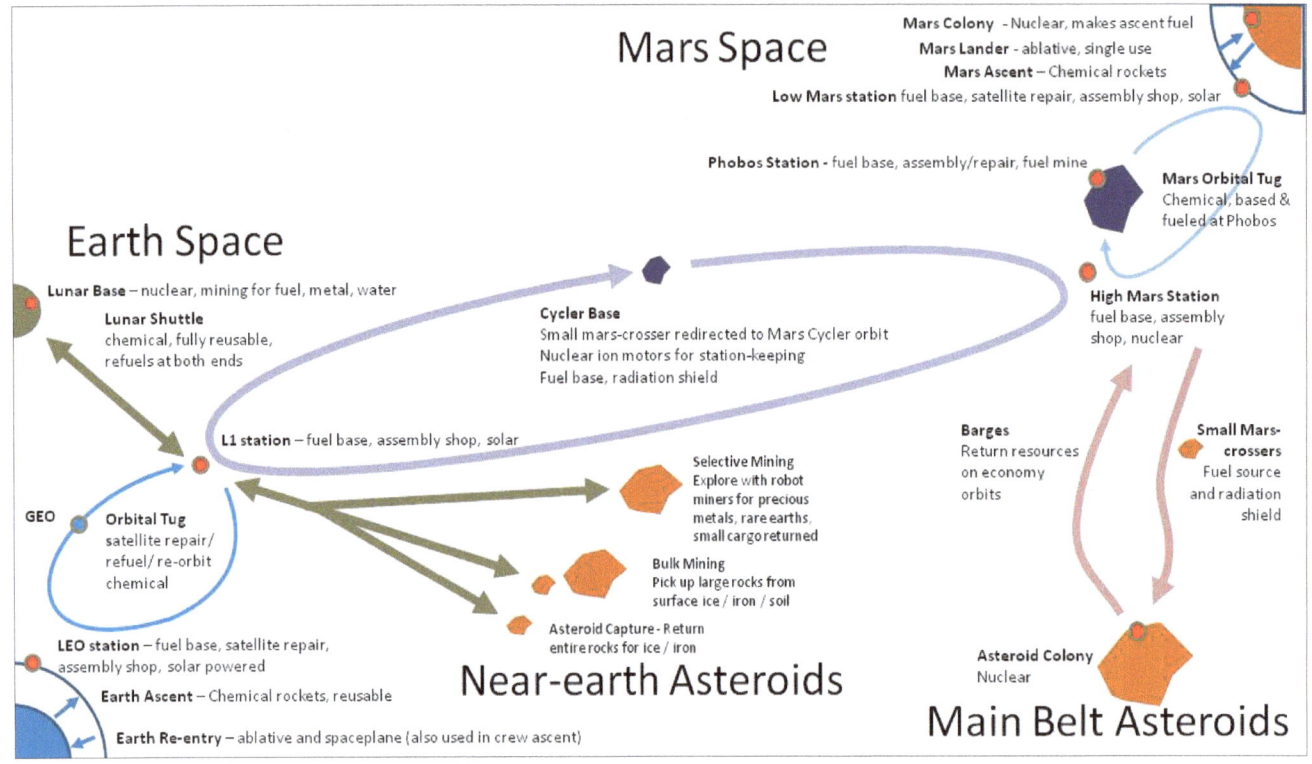

9.7 Timeline

2031	Satellite repair and salvage business, initially in **LEO**, with robot and crewed tugs travelling out to GEO and the graveyard orbit. *Spacetug Copenhagen.*
2033	*Pharos* station moves to L1, and mining of the **Moon** begins in several places, mostly to supply space industries. Some uncrewed mining of Near-Earth Asteroids. *Caverns of Procellarum.*
2035	First crewed **Near-Earth Asteroid** expedition, with ES *Shoemaker* setting out from *Pharos*. *Mutiny Near Earth.*
2037	ES *Shoemaker* returns from the asteroid and slings on to Mars; first permanent settlements on **Phobos**, **Deimos** and **Mars** itself.
2045	The first large transfer of colonists to the **Belt**, using the asteroid **8013 Gordonmoore** as a fuel station and radiation shelter. 2000 people set out, aiming for a dozen asteroids including **81 Terpsichore**. (undefined future book) and **4 Vesta**.
2044	**Phobos** and **Deimos** colonies are well established, and many **Belt** colonies from this time forward start from there rather than from Earth.
2050	*The Dark Colony*
2051	*Freedom at Feronia*
2053	*Traders of Arkady*

9.8 What About the Stars?

In the period I have been studying, even the outer planets of the Solar System are impossible to reach. The nearest star is a hundred thousand times as far away as the asteroid belt. The fastest ships described might take five thousand years to get there.

But, the colonies in the asteroid belt are learning how to be self-sufficient, and how to live together in such a small space. They are the nearest thing to a generation ship we can imagine. The idea is very much in their minds; if you just like to travel for the sake of it, an infinite journey is the next step. The question is not "why go to the stars?" The question is

Why would you ever stop?

Author's Note

Thanks for reading this book. If you have bought the print edition, I will be happy to send you a PDF version free. E-mail me at dickpenn@gmail.com. If you plan to copy pictures or parts of the book, you are welcome, though I would ask you to include my name.

Please do not distribute the book as a whole in electronic form; let me know who you want to send it to, and I will e-mail a copy.

I would love to hear from anyone who knows where I have gone wrong, or who has ideas they would like to see in a future revision of the document. Please contact me on Twitter @RichardFPenn, on Facebook in the group Asteroid Police, or through my website at lockhand.org.

And please post a review on Amazon; it is the only way independent authors sell books.

www.ingramcontent.com/pod-product-compliance
Lightning Source LLC
Chambersburg PA
CBHW050734180526
45159CB00003B/1223